SECRET CHAMBERS

Martin Brasier is Professor of Paleobiology at the University of Oxford. His duties have included Chairman of the Faculty of Earth Sciences; Chairman of the Subcommission on Cambrian Stratigraphy (overseeing the formal definition of the Precambrian-Cambrian boundary); and membership of NSF and NASA panels. He is best known for his work on earliest life and the unique lessons that are now being learned from the early fossil record. Recent papers include laser mapping of the Ediacara biota; complex life on land at 1000 Ma; 3430 Ma cellular fossils showing sulphur metabolism; and a pumice hypothesis for the origins of life. His work on the Ediacaran and Cambrian explosion was recounted in *Darwin's Lost World-The hidden history of animal life* (published by Oxford University Press for the Darwin centenary in 2009).

D0123974

MARTIN BRASIER

SECRET CHAMBERS

the inside story of cells
and complex life

OXFORD
UNIVERSITY PRESS

OXFORD
UNIVERSITY PRESS

Great Clarendon Street, Oxford, OX2 6DP,
United Kingdom

Oxford University Press is a department of the University of Oxford.
It furthers the University's objective of excellence in research, scholarship,
and education by publishing worldwide. Oxford is a registered trade mark of
Oxford University Press in the UK and in certain other countries

© Martin Brasier 2012

The moral rights of the author have been asserted

First Edition published in 2012
First published in paperback 2013

Impression: 1

Published in the United States of America by Oxford University Press
198 Madison Avenue, New York, NY 10016, United States of America

British Library Cataloguing in Publication Data

Data available

ISBN 978–0–19–964400–1 (hbk)
ISBN 978–0–19–968349–9 (pbk)

Printed in Great Britain on acid-free paper by
Clays Ltd, St Ives plc

PREFACE

The Lecturer projected a final image onto the screen. It was the concluding slide of his course. For once, though, it was not a reconstruction of some long extinct creature captured during its final agonizing moments. Enigmatically, it showed an old sepia-tint of the Sphinx of Egypt, surrounded by Bedouin, camels, and sand. This infamous hybrid—a chimaera, half cat and half king—was the very picture of painful transition.

'Look closely at this image' said the lecturer, 'and answer me this question: *what was the most painful and difficult transition in the whole history of life on Earth?*' The class fell silent, not wishing to give voice to any signs of ignorance or superstition. For this was an audience of degree-level students, and ignorance could nag at them like toothache. After a few seconds, though, one brave soul spoke up.

'Did it involve Darwin's Lost World?' Quite a good answer, that one. Over previous weeks, the class had been initiated into the mysteries of the Cambrian explosion. They had learned how the rapid appearance of animal life in the fossil record—some 543 million years ago—was dazzlingly abrupt—rather like the Sphinx emerging from the sands of the desert. So abrupt, in fact, that it greatly vexed Charles Darwin and his successors. But the latest evidence from Darwin's Lost World was showing that, while the Cambrian explosion was rather rapid, it was likely to have been a predictable transition, a natural consequence of all that went before.[1]

'No, it was not the Cambrian explosion' replied the lecturer. 'While that event was *almost* the oddest thing that ever happened in evolution, there was arguably an evolutionary change that can seem to us even *odder still . . .'* He peered quizzically across his spectacles, inviting deeper thought from the class.

'Could it have been the origins of life itself?' suggested another. That was quite a good answer too. In this case, the Sphinx emerging from the sand might be seen as symbolizing life emerging from the waters of the primeval soup. But this answer was regarded by the lecturer as rather wide of the mark as well. Recent evidence, he said, was starting to suggest that the early Earth was remarkably conducive to the synthesis of complex organic compounds. If so, then life may have arisen quite early, quite readily, and perhaps many times over.

The lecturer then began to warm to his theme: 'Even odder than the origins of life? What on Earth could it be?' A silence followed, almost like an increment of geological time. This silence he allowed to linger, so as to build up the quizzical tension in the room: '*This transition was so big and so difficult that it arguably took up most of evolutionary time . . .*' Several gaped in disbelief at this news. How could an evolutionary birth be so long and painful? And why did the lecturer say that a little step in evolution, which we shall shortly unearth, was perhaps the most difficult and unlikely thing that has ever taken place in the history of life? Only time will tell. Happily for us, time is the story that Earth tells best.

That lecturer standing at the lectern was of course myself. And this book is based on a course that I have been giving at Oxford University over the last decade. In it, we begin by travelling back to 1665, to find Robert Hooke being puzzled by his discovery that life can be made up from cells. Travelling forward to 1839, we see how the complexity of the modern cell, and of modern reef ecosystems too, was starting to emerge. By 1859, Charles Darwin was drawing attention to his Lost World in evolution—later found to have spanned more than 80 per cent of Earth history. This book tells the story of the quest to understand the complexity of the complex modern cell, and of the quest to rescue its hidden history from deep within the fossil record.

As with *Darwin's Lost World* (Brasier 2009), this book tells the story through my own researches, first embarking on a cruise as Ships' Scientist aboard *HMS Fawn* in 1970, to map out and explore living Caribbean marine ecosystems. Like my own researches, it then pushes ever further backwards through time, through the puzzlement caused by mass extinction events which began some 600 to 500 million years ago, through to the enigmatic Boring Billion. In so doing, we explore early to modern debates about the nature of some of the oldest known complex cells and the fossils they have left behind (Figure 1). Each step in this narrative draws us backwards towards ever more remote and little known parts of the planetary landscape, and towards equally puzzling parts of the human mental landscape too. I have therefore sought, in each chapter, to put some of the major questions into context by descriptions of premier field locations from around the world, enlivened by descriptions of their fossils, their fossil hunters, and their puzzles. My hope is that the book will reveal just how rich and diverse have been our ways of thinking about the earliest life forms, written in words that can hopefully be read with ease and enjoyment.

Nothing in science would be possible without the support of a network of friends and colleagues. Of special value to this book have been those rewarding interactions with my good friend and colleague Lynn Margulis (who sadly died before the publication of this book), as well as Andy Knoll, Tom Goreau, Paul Strother, and Tom Cavalier-Smith, and the late John Lindsay. I here express my deep gratitude to the following fine teachers and mentors for encouraging an involvement in a lifetime's research into early life, roughly from the 1960s onward: John Clark, Paul Morris, Tony Barber, Bill Smith, John Dewey, Martin Glaessner, Perce Allen, Roland Goldring, Marjorie Muir, Stewart McKerrow, Francoise and Max Debrenne, Michael House, John Neale, John Cowie, Peter Cook, John Shergold, Hugh Jenkyns, and Stephen Moorbath. I thank Andrew Gooday, John Murray, and Jan Pawlowski for their encouragement of work on early Foraminifera; Stefan Bengtson, Roger Buick, Graham Budd, Nick Butterfield, Dima Grazdhankin, Emmanuelle Javaux, Soren Jensen, Malgorzata Moczydlowska, and Gonzalo Vidal, Nora Noffke, Malcolm Walter, Charles Wellman, and Rachel Wood for many valuable discussions about early microfossils, microbial mats, and early habitats over the years.

The following geologists from around the world are here lauded for their invaluable assistance during field work, often in remote and hostile places, followed up by laboratory work, over four decades: Owen Green for his support in the field and for running the Palaeobiology Labs at Oxford; my extended family of students and protégés including Duncan McIlroy, Graham Shields, Louise Purton, Gretta McCarron, David Wacey, Jonathan Leather, Zhou Chuanming, Nicola McLoughlin, Katrina Marsden, Jonathan Antcliffe, Maia Schweizer, Richard Callow, Alex Liu, Leila Battison, Latha Menon, Jack Matthews, Sean McMahon, Kate Hibbert, and Tom Hearing, for many years of lively and fruitful discussion in field, lab, and pub.

Many colleagues also provided much valued support and friendship during the fieldwork for this book: Tom Barnard, Alex Smith, Peter Dolan, Peter Wigley, Sally Radford, John Scott, John White, and the captains and crew of HMS *Fawn* and HMS *Fox* during our cruise around the Caribbean in 1970; the staff of the Petroleum Research institute in Cairo helped support work in Egypt during 1982; David Watters, Jack Donahue, plus the Carnegie Institution of Pittsburgh, and the people of Barbuda gave support to field work in Barbuda during 1984; David Wacey, Leila Battison, Alex Liu and Sean McMahon helped with fieldwork in the Gunflint chert during 2009 and 2011; Duncan McIlroy provided major help with fieldwork in Canada since 1987; John Hanchar and his team at Memorial University in Newfoundland are also thanked for their encouragement and strong support. John Lindsay, Cris Stoakes, Martin van Kranendonk, Kath Grey, Arthur Hickman, Nicola McLoughlin, David Wacey, and my son Alex Brasier are thanked for their great enthusiasm, encouragement, and valued companionship during fieldwork across Australia at various times from 1998 to 2006; the Royal Society for their support of field work in Barbuda, Cuba, Egypt, and Australia; St Edmund Hall, Oxford University, and the Natural Environments Research Council for assistance with travel fieldwork costs over many

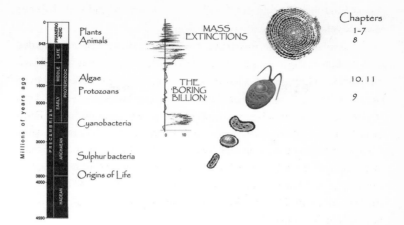

Figure 1. Timelines. As this geological chart shows, the Precambrian encompasses more than 80 per cent of earth history (shown in black), in contrast with the younger Phanerozoic spanning the last 543 million years (shown in white). When Robert Hooke published his first report of living cells in 1665 (Chapter 1), he also predicted that cells could be found as fossils. But it took a further two centuries to discover that all of life is cellular (Chapters 1 and 2), and yet another century to realise that the bulk of geological time has mainly been consumed with cellular innovations. Chapters 2–6 explore Caribbean reefal symbioses and their many ups and down, as a context for the symbiotic origins of the eukaryote cell (Chapter 7). Chapter 8 explores the regular collapse of reefal symbioses back through geological time. Chapter 9 digs into the 1900 million year old Gunflint chert, to a world before the first complex cells. Chapter 10 explores the earliest eukaryote cells, and their context of the Boring Billion, in a world before mass extinctions. The wiggles in carbon isotopes show this stable interval bracketed between big oscillations 2400 and 700 million years ago. Chapter 11 attempts to bring both mass extinctions and cellular innovations into the widest possible context, and uses the fossil record to make predictions about the future.

decades. There are, alas, very many other friends, colleagues, and early teachers not mentioned by name in this list, and I sincerely hope they will accept my thanks for their undoubted contributions here. I also express my grateful thanks to Latha Menon for acting as a major catalyst for this account and for giving much appreciated advice upon the written word.

Last of all, I thank my parents and my elder brother Clive for helping to imbue in me a great love for nature, the planet, and the past beneath my feet; my wife Cecilia for unstinting help with fieldwork over three decades, and for being a perfect 'geology widow' for months each year; and our children Matthew, Alex, and Zoë for their lively wit and enthusiasm during many happy holidays spent in Wales, Scotland, and France. Both Matthew and Alex have contributed hugely to my thoughts and models towards end of this book. This book would not have been possible without their wit and encouragement.

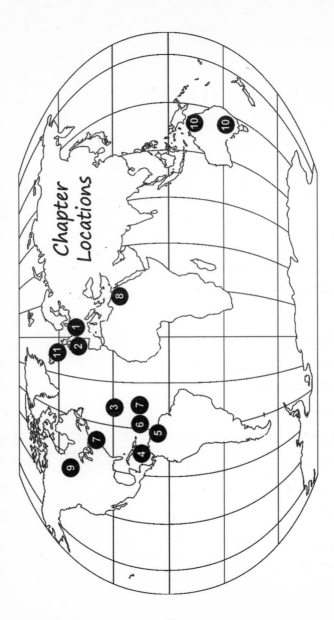

Chapter Locations

CONTENTS

LIST OF ILLUSTRATIONS

LIST OF PLATES

LIST OF ABBREVIATIONS

ADP	adenosine diphosphate
AGSO	Australian Geological Survey Organization
ATP	adenosine triphosphate
DNA	deoxyribonucleic acid
ECOP	European Congress of Protistology
MinLOC	minimum lines of communication
NASA	National Aeronautics and Space Administration
PI	Parsimony Index
RNA	ribonucleic acid
rRNA	ribosomal RNA
SET	Serial Endosymbiotic Theory
SSU	small subunit ribosomal
UV	ultraviolet

CHAPTER 1

THE SECRET CHAMBERS OF ROBERT HOOKE

Plague year

It is September 1665; King Charles and his court have fled London. Left behind are many anxious souls, including the city Aldermen, some apothecaries, a few doctors and a host of grave diggers. Smoke curls through the chill morning air, assaulting our nostrils with a tang of brimstone mixed with animal dung and burning flesh. Outside in the street, a night soil man clacks slowly along the cobbles in worn out clogs. Through a leaded windowpane inside old Gresham College, we can hear his muffled cries, urging us to pile dead citizens on his cart for delivery to the plague pit.[1] Markets are emptying. Shops are shuttered. And the bodies of city dwellers are closing down too—the life of old London seems to be on the brink of collapse. The Apocalypse, it seems, has arrived.

Where has this death-dealing pestilence come from and how does it travel? Superstition is rife across town. Some blame it on a great stench that rises up from the River Thames. Others believe it has arrived with a witch's curse. Citizens are therefore searching anxiously for signs of its progress—little white bumps that first appear in the armpit or the groin. Each discovery is accompanied by a wail of distress. For these—the buboes of bubonic plague—will often be followed by fever, vomiting and death.

All manner of remedies are being attempted to cheat the graveyard of a further crop: a pipe of pungent tobacco, a room filled with wood smoke, the plucking of rosaries and of course, a pocket full of posies. During this time of mayhem, the Lord Mayor has taken to chairing his meetings from within a specially constructed cell—a glazed box with but a single concealed aperture. Curiously, it seems to work—the pestilence cannot travel through his panes of glass.

Solomon's House

Let us now step further inside Gresham College, the first building ever designed for the pursuit of science. Within these walls lives a community dedicated to 'the finding out of the true nature of all things'—a veritable Solomon's House.[2] The college itself has a pleasingly rational construction, being arranged around a courtyard like a Roman villa. Its rooms gain their light from large windows facing the street outside or from the large and airy quadrangle within. The several floors of the college are logically interconnected, too, by staircases that spiral upwards from each of the four corners.[3]

Although rooms and corridors inside this building have been emptied by the pestilence, a whiff of pipe tobacco together with a hint of salted pork betrays the presence of people here just a few days ago. We sneak up a spiral staircase so as to reach our destination, an oak-clad library which sits empty and aloof, near to the northeast corner of the building and facing the great inner courtyard.

Gradually, our eyes grow accustomed to the shuttered darkness here. Around the wainscot, and set between rows of calf-bound books, there sits a teasing cabinet of curiosities—a mouldering human skull covered in lichen; a portion of mammoth tusk, now cracked and grey; and a slab of rock from the Great Pyramid, all covered in stony discs—whorled fossils shaped like the great wheel of life itself. Clearly, this is no ordinary chamber of secrets.[4]

On a table near the fireplace stands a dusty flagon of port, only half drunk. We half expect to see a slice of cheese and a lemon here, perhaps surrounded by curled shavings of peel, as in some old Dutch still life. But there is no cheese and no lemon, only a wine-stained cork, plus a few shavings. How strange to find a sliced up bottle cork. Why has the cork been shaved like that?

Further clues can be found in a leather-bound folio set beside the window. Opening the shutters to let in more light, we start to peruse its pages (see Plate 1). Some have engravings. A few are a little disturbing—hideous creatures furnished with hairy scales, fangs, and spines. Only then do we spy something familiar as we turn yet another page—a picture of that wine cork, or rather, of the shavings obtained from it. It is then that we begin to suspect that this cork has been the object of a remarkable enquiry.

Ever curious, we explore a second chamber hung with mirrors, to draw light into its darker recesses. A wooden lathe stands by the wall here, provided with pulleys of brass and drive ropes of hemp. There is neither electricity nor gasoline to drive this curious old lathe, of course. Instead, the source of power has been a pair of human legs beating away on a treadle board. Closer inspection shows us how this lathe has been used for lapping discs of glass. Piles of glass chippings lie heaped upon the floor like sawdust. In the centre of this chamber stands a bench like a butcher's block. On it, we see further fruits of this labour—twinkling piles of lenses, dozens of them, varying from fat to thin, and from convex to concave. For this is not a Hall of Mirrors, like that being assembled across the Channel at Versailles near Paris. This is a Hall of Lenses. Perhaps the most significant pile of lenses ever assembled.

At last, the penny drops. The year is 1665 and the Apocalypse has arrived—in this very room. That book we saw on the table (Plate 1) was called *Micrographia*.[5] It is not the work of any Old Master. The chamber, with all its books and bones, is no ordinary crime scene. This workshop, which has lain shuttered and forlorn for so many months during the Great Plague, is not a chamber of secrets. It is the haunt of a visionary scientist called Robert Hooke. In it have been gathered together some of the best clues to the nature of life on Earth, using a pile of lenses made by his own ingenious machine.[6] The Apocalypse has truly arrived. This word—in Greek—means *the arrival of the time of truth*.

The hall of lenses

Robert Hooke describes his chamber like this: 'I make choice of some Room that has only one window facing South, and at about three or four foot distance from this Window, on a Table, I place my Microscope, and then so place either a round Globe of Water, or a very deep plano convex Glass…that there is a great quantity of Rayes collected and thrown upon the object…'.[7] In this chamber, he had been paring life back to its barest essentials. That is, until some of those essentials—cells of the bacterium *Yersinia pestis*—had come knocking at his door and he was obliged to flee for the country.

Although rather ordinary in outward appearance, Hooke was the possessor of no ordinary mind.[8] Mercurial, subtle, versatile, and dynamic, he was a scientist of truly marvellous invention.[9] Together with a wide circle of contacts, he was trying to map out how a modern scientific world might begin to work in practice, following some of the precepts laid down some fifty years before by Sir Francis Bacon.[10] Using a mixture of chemistry, curiosity, and craftsmanship, Hooke had been demonstrating inventions to the Royal Society that ranged from watches and optical devices for safer navigation on the one hand, to gadgets leading towards the greater happiness of humanity on the other.[11] Not least among his many contraptions was that combination of glass lenses we have just encountered, which together forms the basis of the modern compound microscope.[12] Along the way, he made use of a surprising spectrum of materials for lenses, as he tells us with delightfully quaint spellings: 'I have made another [microscope] of Waters,

Gums, Resins, Salts, Arsenicks, Oyls, and with divers other mixtures of watery and oyly Liquors. And indeed the subject is capable of a great variety; but I find generally none more useful than that which is made with two Glasses, such as I have already described.'[13]

In this manner, Hooke was able to prepare a hugely magnified drawing of a human flea as seen down his microscope (Plate 2a); as well as a monstrous louse that still clings to a human hair, perhaps his own: 'my little Objects are to be compar'd to the greater and more beautiful Works of Nature, A Flea, A Mite, a Gnat to an Horse, an Elephant, or a Lyon.'[14] Large portraits of these tiny parasites were then engraved by artists—ranging from very good to indifferent—onto sheets of copper, inked up and then sent out into the world from a printing press set up at the Bell Inn near St Paul's Churchyard (see Figure 2).[15] It is easy for us to forget the astonishment with which these portraits were received, not least by a city in the grip of the Great Plague.[16]

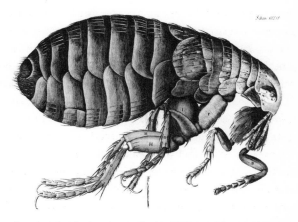

Figure 2. The remarkable engraving of a flea, first published in 1665 from drawings made down the microscope by Robert Hooke. The flea was likened by him to a mediaeval knight dressed in a suit of highly polished black armour, adorned with porcupine quills and steel bodkins.

Source: See Hooke 1665. Observ. LIII. *Of a* Flea.

© A Flea from Microscope Observation by Robert Hooke (1635–1703), 1665 (engraving) (b/w photo), English School, (17th century) / Private Collection / The Bridgeman Art Library

A lovely green

Deep within the pages of *Micrographia* can be found a seminal clue for the story that follows. While studying water in a rain butt, Hooke had watched with fascination as it started to turn into something resembling green pea soup. Scraping away at the scum and examining it with his little lenses, Hooke was amazed to find it filled with tiny moss-like plants. '…this Water would, with a little standing, tarnish and cover all about the sides of the Glass that lay under water, with a lovely green; but though I have often endeavour'd to discover with my *Microscope* whether this green were like Moss, or long striped Sea-weed, or any other peculiar form, yet so ill and imperfect are our *Microscopes*, that I could not certainly discriminate any.'[17]

More to the point, the blooming of life in a water butt caused Hooke to wonder aloud about that deepest of all questions—where does life come from? Does it only come from other forms of life? Or can it be generated 'spontaneously'—from nothing more than air, water, and putrifying organic matter.[18] Hooke was deeply puzzled by this question. In the end, he came to the conclusion that life was being created from inert materials, right beneath his very nose. His reasons for arguing this were not at all mystical. They were entirely logical. Wherever life was known to beget other life, it always did so by means of 'seeds' of propagation. This was nothing more than commonplace, farmyard knowledge. If one wanted a year's supply of mutton and mashed turnips, then the ram must be allowed to inseminate the ewe, and the turnips must be allowed to flower and produce seed. That is what spring is all about.

But look as he might, Hooke could find no signs of seeds within the green moulds or mushrooms he was studying down the microscope. We now know that such seeds—called spores—do indeed exist, but Hooke was unable to resolve them using his primitive microscope.[19] From this lack of evidence for 'seeds', Hooke felt obliged to regard his newly discovered world of mosses, moulds, and algae as part of a highly mysterious, shape-shifting domain. This domain, he thought, lay somewhere between the kingdom of minerals on the one hand and the kingdom of plants on the other. He was mistaken in this view, of course. Mosses and algae really are types of life akin to plants. But while much later scientific experiments were to prove Hooke's ideas about spontaneous generation very wide of the mark, his conclusions should not be ridiculed.[20] Spontaneous generation was arguably a reasonable hypothesis for the time given the lack of knowledge and evidence. Nor should Hooke's ideas be regarded as bad science. The latter, remember, consists of weak ideas poorly tested. But Hooke's hypothesis of spontaneous generation of life in a water butt was eminently testable. And tested it was—with Hooke's own discoveries.

Of dragons and dungeons

Hooke's microscope revealed an amazing amount of detail within the living world.[21] The flea was likened by him to a medieval knight dressed in a suit of highly polished black armour, adorned with porcupine quills and steel bodkins. Grand vector of the Bubonic Plague, it was literally dressed to kill. Not only that, but he noted that '...it has a small *proboscis*, or *probe*...that seems to consist of a tube...and a tongue or sucker... which I have perceiv'd him to slip in and out....with these Instruments does this little busie Creature bite and pierce the skin, and suck out the blood of an Animal, leaving the skin inflamed with a small round *red spot*.'[22] This red spot—the 'Ring around of Roses' of the nursery rhyme—would have been the first visible symptom of a flea bite, leading to invasion by the dreaded plague bacterium.

For the frequenters of coffee houses in early Restoration London, this greatly magnified world of flies and fleas did not at first present any worries. All that was to come later. Rather, it revealed a planet furnished with an almost infinite richness of form. The writer Dean Swift was later to encapsulate this in his famous epigram:

> Bigger fleas have smaller fleas
> Upon their backs to bite 'em.
> And smaller fleas have lesser fleas.
> And so *ad infinitem*.

Even so, as has been hinted, it is not merely for this prospect of an endless succession of fleas that Hooke deserves our plaudits. It is for his detailed depiction of cork tree bark. Now cork might seem rather prosaic in comparison with a flea. But the cork tree has great dignity—it is a kind of oak tree, akin to timbers forming the ties of cathedral roofs, or the hulls of naval frigates. And the bark from one old Cork tree in particular was to perform a mighty magic on the mind of man:

> I took a good clear piece of Cork, and with a Pen-knife sharpen'd as keen as a Razor, [and] cut a piece of it off, and...then examining it very diligently with a *Microscope*...[and] with a deep *plano-convex Glass*, I could...perceive it to be all perforated and porous, much like a Honey-comb, but that the pores of it were not regular...[and that the] walls (as I may so call them) or partitions of those pores were neer as thin in proportion to their pores, as those thin films of Wax in a Honey-comb are to theirs. Next...these pores, or **cells**, were not very deep, but consisted of a great many little Boxes, separated out of one continued long pore, by certain *Diaphragms*.... [These were] the first *microscopical* pores I ever saw, and perhaps, that were ever seen, for I had not met with any Writer or Person, that had made any mention of them before this...I concluded there must be...in a Cubick Inch, above twelve hundred Millions, or 1259712000...a thing almost incredible, did not our *Microscope* assure us of it.[23]

Hooke was the first to record seeing these millions of tiny box-like cubicles, each of similar size and shape, and just a few microns across (Plate 2b). He was the first to call these little boxes 'cells' from the Latin word *cella*—meaning a little chamber that contained a sacred shrine (Figure 3). Each and every cell is like a secret chamber because it is surrounded by a protective wall made of fatty phospholipids. But it is also a secret chamber in another sense too, because its existence was entirely unknown until the invention of the microscope.

Petrify'd

In 1665, Robert Hooke did not confine himself to these little chambers held deep within cork tree bark. He let his curiosity float free, reporting on the first microscopic studies of fungi and slime, rocks and rust. Both

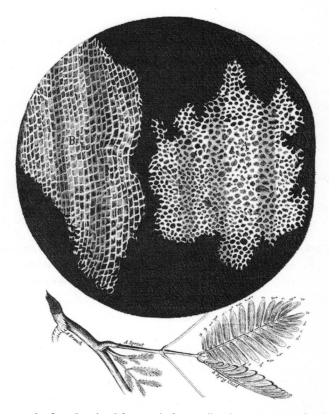

Figure 3. The first clue that life is made from cells. This engraving of cork tree bark in 1665, was seen down the microscope by Robert Hooke.

Source: See Hooke 1665, Observ. XVIII. *Of the* Schematisme *or* Texture *of* Cork, *and of the Cells and Pores of some other such frothy Bodies.*

charcoal and coal were found by him to be constructed from cells: 'a [charcoal] Stick of an Inch Diameter, may containe no less then seven hundred and twenty five thousand, besides 5 Millions of pores [cells], which would, I doubt not, seem even incredible, were not every one left to believe his own eyes ... I found it to look very like an Honey-comb ...'. And later on in the same work, he writes the following prescient words: 'I rather suspect ... [that coal] was at first certain great Trees of Fir or Pine, which by some Earthquake, or other casualty, came to be buried under the Earth, and was there, after a long time's residence ... either rotted and turn'd into a kind of Clay, or *petrify'd* and turn'd into a kind of Stone, or else had its pores fill'd with certain Mineral juices, which being stay'd in them, and in tract of time coagulated.'[24] In this way, Hooke therefore provided us with the first ever reports of fossilized cells preserved in rocks. He conjectured on the possibility that coiled fossil ammonite shells from the coast of England could once have been living creatures.[25] And he even conducted for us the earliest experiments on fossilization, including the petrifaction of cells in wood.[26]

Little wheel of life

While examining a parcel of sand from the Thames estuary in the early 1660s, Robert Hooke found himself peering again at the wheels of life, but these were surprisingly tiny—no bigger than the full stop at the end of this sentence. When viewed down the barrel of his microscope, he was intrigued to find the sediment filled with myriads of miniscule shells: 'I was trying [out] several small and single Magnifying Glasses, and casually viewing a parcel of white Sand, when I perceiv'd one of the grains exactly shap'd and wreath'd like a Shell, but endeavouring to distinguish it with my naked eye, it was so very small, that I was fain again to make use of the Glass to find it; then, whilest I thus look'd on it, with a Pin I separated all the rest of the granules of Sand, and found it afterwards to appear to the naked eye an exceeding small white spot, no bigger than the point of a Pin. Afterwards I view'd it every way with a better *Microscope* and found it on both sides, and edge-ways, to resemble the Shell of a small Water-Snail with a flat spiral Shell: it had twelve wreathings [chambers]…all very proportionally growing one less than another toward the middle or center of the Shell, where there was a very small round white spot. I could not certainly discover whether the Shell were hollow or not, but it seem'd fill'd with somewhat, and 'tis probable that it might be *petrify'd* [fossilized] as other larger Shels often are.'[27]

The shell he illustrated in his first plate of engravings, is now called *Ammonia*. Its chalky shell—properly called a 'test'—is made from dozens of little chambers within which hides a single, amoeba-like cell (Figure 4). Each chamber is interconnected by a small door-like opening, called a foramen, after the word for a hole in the ground in Latin. These little wheels of life are properly called Foraminifera—or, more usually, 'forams' for short. While they resemble tiny fossil ammonite shells, of the kind familiar to us from Jurassic and Cretaceous rocks, they have not been made by extinct but squid-like molluscs. They are made by single celled protozoans, rather like an *Amoeba* provided with fine thread-like pseudopods. They abound in the seas and oceans of the world (Plate 3). They can

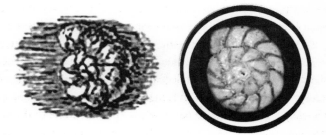

Figure 4. Hooke's drawing of a microscopic, single-celled Foraminiferan from the River Thames (at left), now called *Ammonia*—compared with an example seen down a modern microscope (at right). Its tiny shell is just a fraction of a millimetre across, while his drawing was about one centimetre wide. It was the first image of a single-celled organism.

Source: See Hooke, Micrographia, Observation XI. *Of Figures observ'd in small Sand.*

also been found through much of the fossil record of the oceans. And they are now known to have evolved very fast, so that their changes in shape and size can be used as a tool to date the rocks in which they are found.[28] With this simple description, Robert Hooke had drawn back the curtain on what was to become a powerful tool for our decoding of Earth history.

We can now settle back to enjoy a delicious paradox about this work of Robert Hooke's. Back in 1665, we have seen that London was beset by the Plague, and by this question too—where did the pestilence come from? And with regard to the greening in his water butt—where did such verdant life come from? The answer to both questions was actually to hand. But it was not spontaneous generation, as Hooke had supposed. Instead, it was to come from the marine shells and cells like those he had described for the first time. We can appreciate this paradox now. Unhappily for human society, however, none of this yet fitted together. The microscopic world was allowed to go about its business largely unseen.

Shoes of giants

Being practical, versatile, and well ahead of his time, Hooke might have seemed destined for great honours. But practicality in science is commonly trumped by theory, while versatility is readily bested by narrowness. And while old Hooke was rather lucky with his experiments, it has to be said that he was less than lucky with his adversaries. One of his detractors, it seems, was Sir Robert Moray, a one-time spy for France. An even more powerful enemy was Mr Henry Oldenburg, the Secretary to the Royal Society—later imprisoned in the Tower of London as a spy 'for writing news to a virtuoso in France, with whom he constantly corresponds in philosophical matters'.[29] Those secret messages may well have contained reports about Robert Hooke's early experiments with the microscope.

Not least among Hooke's detractors, of course, was that growing luminary called Sir Isaac Newton. It was Newton, remember, who was famous for 'standing on the shoulders of giants'. He was fabled, too, not only for his optics and his gravity, but also for his self-importance and jealousy. So Newton was not a safe man to cross at any time. He could fiercely flex his intellectual muscles. Worse still, Hooke had developed an unfortunate knack for *stepping on the shoes of giants*. In this way, he unwittingly unleashed the beast in Newton.[30] That may explain why Newton later tried to crush both Robert Hooke and his legacy.[31]

Today, we need to twiddle with the focus knob in order to see Robert Hooke in fuller perspective. Yes indeed, he was outwardly rather unkempt—something with which many of us can easily identify—and he could be disagreeable too. But he evidently helped our ancestors to unveil a new and important kind of vision—of a riotously seething microcosm that lies just beyond our optical reach. This novel viewpoint—often called bottom-up thinking–sees the world as though through a convex lens. It raises the significance of very little things, revealing as much activity at the microscopic level as can be seen at the macropscopic level. Indeed, bottom-up thinking can go even further, suggesting that we are deeply dependent upon the activities of microscopic forms of life.[32] For some of my colleagues, it seemingly points towards a kind of Microtopian vision, in which it is microbes that really rule the world.

To those more at ease with looking at the world through a diminishing lens—with order flowing down from God to man to microbe—this Microtopian vision can sound very unsettling. That is because it turns the cosy logic of such top-down thinking on its head. Worse still, as we shall later see, bottom-up thinking can lead to the heretical thought that big and complex systems are more prone to collapse than are little systems. Big systems might contain the seeds of their own destruction.

It also needs to be admitted that Hooke's legacy was very far from assured in the plague year of 1665. His chamber and collections at Gresham College only survived the Great Fire of 1666 by a whisker.[33] And in the following decades, it was not so much the compound device of Robert Hooke that captured the headlines but his own bobblescope, a somewhat cruder device. The latter was able to provide a much easier way of making a high-powered lens. That is because the power of a lens gets greater as its diameter gets smaller. So all that was needed was a tiny bobble of glass. And that is exactly what happened a decade later, when a Dutch linen draper called Antonie van Leeuwenhoek began to resolve not only the presence of cells but also cell motions, cell contents, and even bacterial cells.[34]

But even this charming little bobblescope was found to have failings, too. The observer was barely able to change its focal depth or correct for optical distortion.[35] Sad to report, there followed a 'technical hitch' in the development of bottom-up vision that lasted for decades. Grinding of lenses was found to be both time consuming and difficult. After dozens of hours in grinding and dozens of hours in polishing, there could still be flaws in the glass, or flagrant distortions along the edges of the field of view. No respectable scientist wanted mockery for working with something little better than a distorting mirror at a country Goose Fair. Their resolving power remained poor. Indeed, the microscope was therefore at risk of becoming little better than a sideshow—until about the time of the French Revolution. This turbulent period, lasting several years from 1789, heralded yet another way of re-ordering things both small and large.

By 1830, many of the more annoying distortions and aberrations of the microscope had been resolved, thanks to a wine merchant named Joseph Lister.[36] And by the start of Queen Victoria's reign in 1837, naturalists were starting to discover that cells were not confined to cork trees or charcoal or even coal plants at all. They could be found in every kind of living matter. Even in kings and queens themselves. Nearly all of life on Earth—well over 99.999 per cent in terms of bulk—was seen to consist of single cells alone. Such a dangerous conclusion was given a great boost by a German physiologist called Theodor Schwann and a German lawyer called Matthias Schleiden, who studied plants for a hobby.

In 1838, according to a famous legend, Schleiden and Schwann were enjoying after-dinner coffee together and talking about their studies on cells,[37] only to discover they were finding similar things in both plants and animals. The two scientists went immediately to Schwann's lab to look at his slides. Schleiden had been studying plant cells. But Schwann had devised a staining technique for studying animal cells, which are much harder to see. The latter then published a book on animal and plant cells in 1839, strangely leaving out any acknowledgement to Schleiden.[38] A cell, said Schwann, was the common currency of life. Its presence might therefore be used to define life itself, a notion called Cell Theory.[39]

This was no very ordinary or expected conclusion. It was something very extraordinary indeed. So strange, in fact, that it was soon pounding away at the very foundations of another great barrier—that between ourselves and the rest of the living world. If all living things are cellular, for example, how does a colony of cells turn itself into an individual blessed with a soul?[40] That was an unsettling thought on its own. But it was becoming possible to conceive of an equally worrisome notion, too: that familiar things from babies to baboons, and from bananas to bacteria—including plague bacteria—may all share in a deep and seemingly hidden history; the hidden history of the cell.

The microscope was therefore poised to reveal some new and startling things about the world in which we live—or the world in which *we think* we live. To explore this mystery further, it is time to dress up in our finery, rattle across old London, and hopefully arrive in time for dinner in Bloomsbury.

CHAPTER 2

A DINNER IN GOWER STREET

Reception room

One evening in April 1839, several horse-drawn cabs could be seen drawing up outside a house in Gower Street. Out stepped two of the world's most eminent scientists. Both appeared uncertain as they approached the steps of Number 12 for the first time—for this was the new home of Charles Darwin, recently returned from his five years at sea aboard *HMS Beagle*.[1]

After ringing the doorbell, the guests were ushered upstairs, and the servants scurried about below. For this building—like many a London merchant's house—was markedly hierarchical in structure. Built of ugly yellow bricks that were blackening in the smog,[2] it formed part of a long chain of tall thin homes, each four stories high. Its interior design was little better than a pile of shoe boxes. Servants either worked in the darker basement or slept among the icy top floor chambers, scurrying up and down via a narrow back staircase. Darwin and his family occupied the middle three floors, connected by the much grander front staircase up which our visitors have just climbed.

While the soup was being simmered in the basement kitchen, Darwin will no doubt have regaled his new visitors—Mr Charles Lyell and Mr Robert Brown[3]—with the latest news upstairs. He showed them how, in this house replete with books and bugs, rocks and rugs, he had provided himself with a study for working in (in the morning), a long back garden for walking down (in the afternoon) and, best of all, a newly married wife for talking with (in the evening). Clearly, all the props for a well-ordered life were now to hand.

The guests will surely have rejoiced at all of this. But they may have quietly recoiled at the colour scheme of his new reception room. Darwin's landlord had installed bright yellow curtains, furniture of red plush, and walls of a gaudy blue. Darwin sarcastically christened this new home 'Macaw Cottage'. The guests will have also noticed that his young wife Emma was due to give birth to a child, the first of a large but comfortable family. But few will have realized that this home-loving naturalist was due to give birth to a theory, an expansive but uncomfortable philosophy on evolution.[4] Darwin's 'dangerous idea', like Emma's baby, was a project in gestation.

It was with the *order* of the natural world at different scales that these three scientist detectives—Darwin, Brown, and Lyell—were grappling in that spring of 1839. Between them, they were contemplating the deeper meaning of mountains and monads. Taken individually, their questions seemed monstrous too. Like many of the big questions in science, their puzzles sounded bizarre—more like riddles than reasoning. For example, Darwin was wracking his brains about *rocks that seemed afraid of the dark*. Brown was trying to get to grips with *cells that have taken in strange lodgers*. Most puzzling of all, Lyell was battling with *worlds that seemed to drop dead for no good reason*. To see where these riddles might yet lead us, it is time to start eavesdropping over dinner.

A puzzle in polyps

First to lean forward to speak at our dinner party in Gower Street, let us say, was the host Charles Darwin. In 1839, he had just finished putting the final touches to his first book, called the *Journal of Researches*.[5] More importantly for our story, he was also in the process of piecing together his first major idea about rocks and organisms. It did not mention earthworms; and it was not about barnacles—both of which studies were to emerge much later. It was about rock-forming creatures called corals. A puzzle in polyps.[6]

During his voyage around the Pacific and Indian oceans, he had been fascinated to see how coral polyps—'coral insects' as he called them—could construct great rims of chalky material around volcanic islands, making them hazardous to approach by sea.[7] Some of these constructions were as low as a row of terraced houses, like those springing up all around Gower Street. Others were as spiky and spooky as any Hawksmoor church in the city nearby. And just like the suburbs of Victorian London, with its flower girls and chimney sweeps, those tropical coral reefs teamed with diversity and colour, replete with sharks and snappers. But how was it possible, wondered Darwin, for any organism as insignificant as a coral polyp to toil away at building such a vast edifice of rock? Did tiny corals really build up their constructions—like those of Cocos Keeling atoll—from hundreds, or even thousands, of feet down on the sea bed? Even great engineers like Mr Marc Brunel had not managed to build anything like that.

Happily, Darwin had come up with an intriguing solution to this puzzle. His idea was at once imaginative in scope, powerful in its predictions, and testable in the rocks and oceans around the world. We need to remember that Darwin was not grappling here with the evolution of species. Not yet. He was thinking about the evolution of landscapes and their ecosystems—bringing together two of his greatest passions—field geology and biology. He suspected that the crust of the Earth was not static at all but in a state of motion. And that coral reefs were only able to grow in shallow tropical waters.[8]

If coral rocks were largely built by corals, as he suspected, and if reef-building corals could not live below 25 feet deep—as he observed on the *Beagle*—then his coral rocks were effectively 'afraid of the dark'. And if they could only grow in the light, then the shape of a coral reef would be hugely affected by any subsidence of the rock on which it was built. Reefs would be forced to grow upwards as fast as their bedrock sank downwards. Hence, a young island with a stupendous cone of rising rocks might show coral reefs that still hugged the shore—fringing reefs, like those of Mauritius. Those islands with old and sinking volcanoes would find its reefs being forced to grow ever upwards, leaving a navigable shipping channel between the reef and the beach—a so-called barrier reef, much as around Tahiti. Lastly, those islands in which all traces of an old volcano had sunk deep beneath the ocean should leave behind something called an atoll, like that of Cocos Keeling—a great circlet of coral separating a deep interior lagoon from the fore-reef beyond. Darwin's studies of naval depth-soundings[9] confirmed that these island slopes dropped rapidly downwards towards the deep, sometimes at a hellish angle.[10] As Darwin himself was to put it: 'On this view every difficulty vanishes: fringing-reefs are thus converted in to barrier-reefs; and barrier-reefs, when encircling islands, are thus converted into atolls, the instant the last pinnacle of land sinks beneath the surface of the ocean'.[11] If he was right on this conjecture, then drillings through the rocks of an atoll should one day reveal volcanic basement deep beneath the surface.

We need to remember that Charles Darwin was not much of a biologist, back in 1839. He was first and foremost a great geologist and a naturalist. The bulk of his notes on the *Beagle* voyage were geological.[12] His early publications were geological. For geology was then the Queen of Sciences.[13] This geological know-how was likewise fundamental to the birth of life sciences. And his writings stayed geological in style to the very end. Without Darwin and geology, biology was arguably about little more than bug collecting. But as soon as scientists added together the ingredients of Darwin + geology, then biology was truly able to go exploring in deep time.[14]

Here, then, came Darwin's big question of 1839. Could big and complex things, like reefs, have tiny and unthinking causes, like polyps? More importantly, could big and complex things *in general* have little and unthinking causes, and be organized from the bottom upwards? It was a heretical thought.[15] And it was a typically geological one. But as we have said, Darwin was a geologist and a pretty good one to boot. He therefore knew that he needed to make huge intellectual leaps to anticipate the strangeness of the Earth itself.[16]

It occurred to Darwin that the modern oceans could be regarded as old and rather weary landmasses that were now sinking down beneath the waves. In that way, different stages in the evolution of coral islands might be comparable with a series of portraits, perhaps like those seen through the lens of a newly invented camera: from a young midshipman (a volcano)—through a flourishing captain (with an expanding halo of little reefs)—to an old admiral (a large barrier reef)—and thence to a final sagging beneath the waves (to leave behind a vast atoll). If so, then big constructions—like atolls—can have very small causes—like coral polyps seeking to avoid darkness. Nothing more was needed.

What also puzzled his colleagues—and his successors—though, was a related question. Why would corals be afraid of the dark? Or to put it another way, why did reef-building corals need to live in the light? That didn't seem to make any kind of sense at all, not least because corals are animals, not plants. Plants plainly need sunlight. But animals, most surely do not? If Darwin's thesis about sunlit corals turned out to be nonsense, then Darwin's first great theory—his theory of coral reefs—would be dead in the water too.

Could any of this be true? Only time would tell.[17]

The little things in life

The next guest to commune over dinner, let us conjecture, was Robert Brown. Everyone around the table knew he had served aboard HMS *Investigator* as an indomitable Ship's Naturalist, collecting plants from the hinterlands of Australia, around 1801.[18] The bonny *Banksia brownii* was named after him, for instance. But the not-so-bonny Mr Brown had long ago left behind the shipboard world of ropes and spars, to live a quieter life in a world surrounded by shiny brass microscopes, botanical mounts, and etched glass slides. True, he had succeeded in becoming the first Keeper of Botany at the British Museum.[19] Now, however, he was focusing inwards upon those little things in life, the tiny details that squiggled and squirmed in the reflected glow of a lamp set beneath his ground glass lens. Miniature monsters called monads; and little cubicles called cells.

Especially curious in this regard was a very peculiar observation made by Brown. He had discovered that cells usually contained a frothy liquid, called protoplasm, whose particles jiggled about constantly. Particles as small as 1/500th of a millimetre across were seen to jiggle back in 1827, and they haven't stopped jiggling since. It seemed as though his cells were possessed by some mystical force. To his eternal credit, he did not attribute these jigglings to the innate forces of life itself. That is because Brown discovered that nearly everything smaller than a printed dot, both animate and inanimate, when immersed in water, could be observed as in a state of constant movement. As we now know, this motion is caused by the bombardment of atoms. So curious was this finding that it was named after him: it was of course Brownian motion.[20]

But it is Robert Brown's next claim to fame that really draws us in here. Whilst poring over thin slices of orchid tissue as seen down the microscope, he had discovered that each of their cells invariably contained a dark blob in its centre, each shaped like a little nut (Figure 5). Each blob was accordingly named by him 'the nucleus', meaning precisely that—*a little nut* in latin. Now, it is true that these dark blobs had been illustrated within plant cells long before, by Antonie van Leeuwenhoek in 1682 and they had even been noted as regular features of such cells by Franz Bauer in 1802. But it was Brown who designated the name of the nucleus for this organelle, back in 1833.[21]

At first, Brown thought the nucleus was only present in those flowering plants that include orchids, onions, and grasses.[22] That conclusion stood up well enough until, as we have seen, Schwann and his colleagues found that the nucleus is present in the cells of all plants, all animals, all kinds of fungi, and even in every kind of protozoan. That might have been an end to it. But there was a final twist to this story.

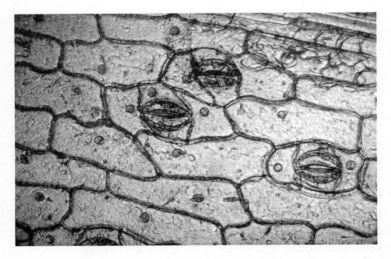

Figure 5. Orchid tissue as seen down Robert Brown's own microscope, taken by Brian Ford. In it can be seen the cell walls, the dark spot-like nuclei, and the lip-shaped stomata. This microscope is now on display in the Linnean Society, London. © Professor Brian J. Ford

By 1858, we have to report that Robert Brown was no more. But even his sad demise made history—it provided a space for the reading of a joint paper on evolution by natural selection, by a certain Charles Darwin and the young Alfred Russel Wallace at the Linnean Society in London. Ironically, it needs to be said that Brown himself might not have approved of this turn of events. For it was Brown who had jilted earlier speculations on evolution by comparing seeds of plants from Egyptian tombs with those still alive today—and finding no measurable difference.[23] Early in 1859, however, the evolutionary cat was slipping out of the bag, and Darwin's *Origin of Species* began to settle down comfortably on the sofas of working men, vicars, and viscounts.

A decade later, Thomas Huxley was starting to draw some interesting conclusions about cell evolution: 'Protoplasm, simple or nucleated, is the formal basis of all life.... But, at the very bottom of the animal scale,... simplicity becomes simplified, and all the phenomena of life are manifested by a particle of protoplasm *without a nucleus.*'[24] Huxley had even started to search for this non-nucleated life. At one time, he thought he had found it too, within dredging obtained by ships such as *HMS Porcupine*. This was the slimy substance he later called *Bathybius haeckeli*. Alas, this slime was not living matter at all but the products of chemical reactions in jars that had festered during storage.[25]

But it wasn't so much Huxley as Haeckel who took this story a stage further. If Thomas Huxley can be called Darwin's English Bulldog, then Ernst Haeckel was his German Shepherd. Like Huxley too, Haeckel was also deeply interested in the emerging science of biology and the mystery of evolution in deep time, writing in 1872 that 'The amoeboid nature of the young egg-cell...justify us in affirming that the oldest ancestors of the human race (as of the whole animal kingdom) were simple amoeboid cells'. So where did that first amoeba come from? Haeckel took a giant leap into the darkness here: 'in the beginning of the organic history of the Earth, at the commencement of the [early Precambrian] period... the Amoebae have originally developed only from the simplest organisms known to us, the Monera [bacteria]. These Monera...are also the simplest conceivable organisms. Their body has no definite form, and is but a particle of primitive slime…a little mass of living albumen, performing all the essential functions of life, and everywhere met with as the material basis of life.'[26]

Huxley and Haeckel were both suggesting that the presence of a nucleus in an *Amoeba* cell and its absence from all bacterial cells could be regarded as the biggest dividing line in the whole of the natural world.[27] If so, said Haeckel, then the earliest signs of life in the fossil record should consist of bacteria alone. This was a courageous prophecy.

Another decade was to pass before scientists were able to show that Robert Hooke's quaint hypothesis about spontaneous generation of cells from minerals, air, and water—like the algae in his water butt—was a wholly mistaken idea.[28] By this time, all living cells could be shown to have originated from other living cells. This bold statement is the basis of what we now call Cell Theory. And Cell Theory may be called 'The First Commandment' of modern biology.[29] It took much longer, though, for scientists to realize that little things such as the acquisition of the nucleus and the chloroplast, those strange lodgers in the cell, may have involved the most difficult and lengthy rite of passage in the whole history of life.

The wheel of fortune

We can now return to our seats at that dinner party in 1839. Let us imagine that the empty platters had been gathered away and the port was passing around the table. Gradually, the hubbub died away, because Mr Charles Lyell was about to speak his mind. Everyone had to strain their ears to catch his words—he spoke so very quietly and horse-drawn carriages were clattering by so loudly outside.[30]

Many around the table were intrigued to hear Lyell's views because he was the author of a major contribution to scientific thinking called the *Principles of Geology*.[31] In this fat tome, Lyell had been seeking to demolish many of the fantastic speculations of eighteenth-century armchair philosophers, men like William Whiston and Louis LeClerc de Buffon with all their periwigged perambulations about catastrophic change. Whiston was the successor to Sir Isaac Newton at Cambridge. He had suggested that extraterrestrial bodies, like those of Halley's comet, played a powerful role in the deep history of life and the planet. Best of all, he came up with the charming idea that the great heat found at depth in the Earth—leading to volcanic activity—was inflamed by the steady accumulation of human sin and lust at its surface. Now, that might sound to us rather like more Bad Science. But it was, at least, an eminently testable hypothesis. Regrettably, there is no evidence that William Whiston ever tested this particular hypothesis with his own experiments.

In place of hellfire and damnation, Lyell was trying to argue that geological change was slow, that the past was vast, and catastrophes were rare. And that sin was not a geological force. Mountains, instead, were blasted flat by things like growing grass. It was a largely bottom-up vision, like that of his protégé Charles Darwin. And it was this line of thought that he liked to advocate with the cool and winning intellect of a Scottish lawyer—which is exactly what he was.

Lyell was perfectly happy to give way to his disciple, Charles Darwin, on the question of coral reefs by this dinner party in 1839. He could draw some satisfaction from the fact that his own pet theory—that the circle of corals in an atoll was caused merely by growth around the rim of an old volcanic crater—had once caused a stir.[32] Indeed, it was one of the great puzzles that the Admiralty wished resolved by Mr Darwin and Captain Fitzroy aboard HMS Beagle. After all, it would plainly be foolish to build a harbour inside the vent of a slumbering volcano. It would be mournful to repeat the terrible mistakes of Pompeii and Santorini.[33] A gentleman needed to know these things when travelling across mare incognita.

Lyell was so excited on his first hearing of Darwin's own theory of coral reefs that he is said to have jumped up and down with joy. But Lyell still had reasons to be glum. For what really vexed him at this time was not the origin of coral reefs. It was the much graver matter of mass extinctions. A French naturalist called Georges Cuvier, like Lamarck before him, had been arguing that the record of past life in the rocks is full of fits and starts. And he was claiming that these jumps in the story are evidence that the history of life is filled with events like the French Revolution—mass killings followed by fresh beginnings. Big systems had been suffering very big collapses. Evidence was being amassed for nearly a dozen of these revolutions in the rock record. The idea was popular in those countries where radical thinking was fervent, in places like France and Holland. And it had resonance in India too, where the Great Wheel of Life was much revered.[34] But such revolutions were seen as terribly un-English behaviour.

The causes for these great catastrophes, of worlds that dropped down dead remained unclear. And as I shall argue later, they are greatly misunderstood even today. In 1839, most people probably assumed that they provided evidence for an angry God, smiting the planet for its cumulative sins. Such a view was just about acceptable when there was evidence for only a single smiting—taken to be that of Noah's Flood. That would have been the view of William Whiston in Cambridge. And it was still, in the 1830s, the view of William Buckland of Oxford. Unhappily for these gentlemen, however, fresh evidence was starting to emerge for nearly a dozen such smitings, long before there were any signs of man on the Earth. Indeed, a group of Frenchmen—Georges Cuvier, Elie de Beaumont, and Alcide d'Orbigny—were starting to reveal such facts in rocks of Mesozoic age. And a cavalcade of Englishmen—Roderick Murchison and Sydney de la Beche among them—were galavanting across Europe to reveal much the same in even older rocks of Palaeozoic age. There was not just a single wiping out of life; there were many.

All of this must have caused members of the English establishment to feel a bit nonplussed. They struggled to see the purpose of a creator in all of this frenzied 'smiting'. Why were those ammonites, those wheels of life depicted by Robert Hooke, no longer with us? And what did those terrible lizards—the dinosaurs—do to deserve such oblivion? Perhaps the creator had made lots of mistakes that needed to be rectified. But that would imply the creator was a bit of an amateur. Or perhaps 'he' was unwilling to intercede. But if so, then why call 'him' good?[35] Or maybe the creator wasn't really an omnipotent force at all, just an imaginary friend.[36] In which case, a top-down vision might be a wholly false perspective, and another explanation was needed.

And this is what Charles Lyell was trying to develop. He was emboldened to argue that these great revolutions in the history of life were not real events at all. Instead, the geological record was faulty; it was full of gaps. It was these gaps, said Lyell, that gave to the fossil record its unpleasant resemblance to a series of French revolutions.[37] In its place, Lyell therefore preferred to conjure a picture of a past world that was calm rather than calamitous, and in which things changed very gradually over vast periods of time. Through thirty long pages in his *Principles*, he laid out his arguments as to why gradual but not abrupt extinction might be the expected pattern—climates slipping slowly into coolness; lands sliding up or down rather gently; new predators arriving in little dribs and drabs.[38] It all sounded utterly reasonable. In fact, it all sounded a little too reasonable, rather like a government official trying to calm a proletarian mob.[39] In the best traditions of English Law and the British Constitution, his rules for geology were therefore to be about gradual change, not revolutionary mayhem and divine intervention. Until we have accumulated evidence to the contrary, said Lyell, then we should assume that everything in the past has worked in exactly the same way as we see things working now—his famous Principle of Uniformity.

Lyell's grand idea was arguably a necessary step—an unwillingness to accept either negative evidence or any kind of pre-destination in the history of life. It was effectively a null hypothesis against which to test the mounting claims of 'progressive development' in the story. Not so much a denial of history as a denial of progress. It was, and still is, a very strong hypothesis for making powerful predictions. But was it true in this case? Were mass extinctions hysterical misunderstandings about the nature of the fossil record—as Lyell was suggesting—or were they real events?[40] Was it to be mass extinctions; or mass hysteria? Only time would tell. Happily for us, time is the story the Earth tells best.

And so to bed

Finally, the clock struck ten o'clock at night, telling the guests it was time to make their farewells. As the carriages clattered homewards, avoiding the infamous slums of the Rookery,[41] we can imagine the scene left behind at Number 12. The house became awash with movement, like cytoplasm streaming through a cell, here moving from room to room and floor to floor. A great pile of Wedgwood plates, themselves a gift from Emma's own father—Josiah Wedgwood—was hurried down into the basement. Here the scullery maid scrubbed them in waters that rapidly became rancid with the smell, and hence the cell contents, of mutton and mint. Emma hurried up to her writing desk on the third floor, where she scribbled excitedly about the success of her first dinner party.[42] Charles Darwin withdrew to his study, sick from the strain of so much talking. But his spirits were raised after lifting a small piece of rock from his pocket, and turning it about in his hand. As shiny and grey as flint, it was part of a chunk found by him several years before, while climbing high in the Andes. The rock had seemed strangely marked at the time, and so it was sent on to Dr Robert Brown in London for closer examination. Now, it had found its way back to its owner.[43] Peering closely, he marvelled at what Brown had seen—the beautiful preservation of fossilized plant tissues—hacked from a tree trunk that had turned into stone. That on its own would have been marvellous. But this was no ordinary fossil tree trunk. It had been found by Darwin standing with dozens of others, much higher than the modern tree line. Such an outrageous setting demanded a sensible explanation. And Charles Lyell already had one on offer: that the Andes was not just a big pile of stones. It was a dynamic mountain range. The rocks were being heaved upwards by gigantic forces deep within the Earth, lifting lakes, fish, and even fossilized forests high into the sky.

Thus it seemed that cells were constantly jiggling about. Coral reefs were sinking down into the sea. And fossil forests were being pushed skywards. Nothing, thought Darwin as he turned down the gas lamps, ever seemed to stay put.

Nested thoughts

From our backward glancing perspective, we can now see that early scientists such as Hooke and Brown were good at observing and picking out patterns, such as the cell, and the cell nucleus. But we can also see that Victorian scientists, such as Darwin and Lyell were keen to get beyond the shape of patterns in the rocks towards conjectures about process— the ways in which life and the universe have worked in practice.[44] Lyell was arguing that there was no grand pattern in the story of life, whereas Darwin already suspected that his mentor was being led up the garden path here. At the time of this dinner party, in fact, Darwin was already coming to the conclusion that a study of extinct organisms would '*throw more light on the appearance of organic beings on our earth, and their disappearance from it, than any other class of facts*'.[45] Darwin was therefore working his way towards another process that could explain all of biology plus much of geology—that process we now call Darwinian evolution.

It is at this point in the discussion that we should come back to reflect again upon those three curious riddles that were mentioned at the start of the chapter. Darwin, remember, was said to be contemplating *rocks that seemed afraid of the dark*. These were his coral reef rocks, which appeared to need sunlight to grow. But surely only plants need sunlight? Why ever did coral animals need the sun?

Lyell was battling with *worlds that seemed to drop dead for no good reason*. Those worlds, like the Permian and the Cretaceous, had inhabitants that seemed to disappear completely during mass extinction events at their close. But if that was so, why did extinctions happen so regularly? Why, indeed, did they happen at all?

Brown was getting to grips with *cells that have taken in strange lodgers*. These lodgers were the structures we now call organelles, such as the nucleus and the chloroplast. But there was a conundrum here: why do some cells come provided with nuclei (like animals and plants) or chloroplasts (like plants) while others (like bacterial cells) get along without such gadgets? What has been life's little game here?

In 1839, these three riddles were believed to be rather separate, at best. After all, what could they have in common to connect them? But good science demands that we turn things over in our minds, that we be playful with our thoughts, and that we seek out the oddities. In brief, it requires us to explore strange connections. As we shall discover through the course of this book, these three riddles are not really separate things at all. They are curiously intertwined. Both coral reefs and mass extinctions provide the key to unlocking the third, and greatest, mystery: the hidden history of the complex modern cell.

It is time, now, to go out into the world in search of answers. And to learn about the trade of a Ship's Naturalist.

CHAPTER 3

SARGASSO

A painted ship

A wet flapping sound caused me to swing around. Something like a little silver torpedo had slithered onto the deck—another flying fish. After a week spent crossing the Atlantic, yet another flying fish was always a welcome distraction. To pass our long days at sea, we had been pursuing a 'mend-and-make do' routine, painting the ship's funnel with a tin of regulation buff pigment. For we were not part of the Grey Funnel British Navy—the battleship navy—but part of the stylish White Navy, with its white hulls, buff funnels, and hardwood decks—the Hydrographic Division of the Royal Navy. As the day wore on, our arms and necks grew burnt from painting and other tasks upon these decks, built up from fresh tar and old teak. An odd sort of silence settled over our ship, too. All sounds seemed to be swallowed up, within a hot and soporific mist, formed of sea spray mingled with tar and turpentine. To paraphrase the

Ancient Mariner, we not only looked, but we smelt, like a painted ship upon a painted ocean.[1]

I found myself crossing the wide Sargasso Sea a mere six months after graduating in London. For a spell before the cruise, I had actually dug myself in, near to the bottom of Darwin's old back garden in Gower Street, being drawn there not merely by his legend but by the spirit of a Regency sage called Jeremy Bentham.[2] As a philosopher and philanthropist, Bentham was greatly interested in cells and their contents. Only it wasn't the cells of plants. It was the cells of prisoners—he was an ardent reformer of the Regency period. By the time of Darwin's dinner party in Gower Street, he had been dead for some years but, strange to relate, was still sitting patiently in his chair, set within a large mahogany cabinet behind glass doors. Indeed, he can still be seen sitting in his Auto-Icon today, complete with panama hat and cane, settled comfortably within the south cloisters of University College London in Gower Street.[3] Bentham was one of the spiritual fathers of that proudly godless institution. And—at his own request—his remains have been mummified for all to see. On special occasions, Bentham's desiccated corpse is still wheeled along to Council meetings, where he is recorded as 'present but not voting'. According to tradition, if the Council vote is tied, Bentham will always give his vote in favour of the new motion: a radical in death as he was in life.

By the time I had become a graduate student in 1969, that old house at Number 12 Gower Street had been demolished and turned into a Lecture Theatre. It was appropriately termed the Darwin Lecture Theatre.[4] Not only that, but the back garden, where Darwin had once taken his daily perambulations, and first contemplated the turning action of earthworms on soil, had sprouted both libraries and laboratories. One of these labs had become a premier centre for the study of microscopic fossils. It was to here that I was drawn by Tom Barnard who was Professor of Micropalaeontology—which means the study of fossils down the microscope—at University College in Gower Street. Or rather, it was to a prefabricated shed—the Caribbean Hut—that I was drawn, huddled beneath the walls of the famous Flinders Petrie Museum of Egyptology.[5] It may have stood at the bottom of Darwin's old garden. And it may have stood beside fragmental remains of the Sphinx. But for all that, it could seem a dark and inauspicious place.[6]

Auspicious or not, 1969 certainly felt like an exciting time to be a young planetary scientist. Worries about the demise of reefs and rainforests still lay some way in the future. There was, as yet, the whole ocean for us to understand. Maritime research was therefore going at full steam ahead. And beyond Earth, NASA was reaching out for the moon, eventually succeeding by July of that year. The phone call inviting me to join UCL even came while I was contemplating Neil Armstrong and Buzz Aldrin getting to grips with Lunar geology. Those Apollo expeditions would show that the Sea of Tranquility on the Moon was some 4000 million years old. But Tom Barnard was inviting me to reach a much younger sea of tranquility—the modern reefs of the Caribbean Sea. And he was asking me to serve as a Ship's Naturalist on a naval survey cruise.[7] I thought about it—for a minute or two. It would have seemed churlish to refuse.

My main home during 1970 was therefore to be aboard HMS *Fawn*, a Beagle Class survey vessel of the Hydrographic Division, and a sister ship to HMS *Fox*.[8] The aim of our cruise was to prepare charts and reports of coral reefs and banks. Bathymetric surveying would be undertaken by the Royal Navy hydrographers—affectionately known as 'Droggies'. And marine biology and geology would be studied by John Scott, Peter Dolan, and myself, from University College—informally known as 'Boffins'.[9] For some six months before departure, we accordingly assembled a hopeful assortment of skills and equipment of the kind needed for rough work at sea: secchi discs to measure the water clarity; diesel winches to plumb the ocean depths; a Maccareth corer to sample the mangrove swamps; a Van Veen grab to grapple the backreef lagoons; log books to record our labours; and so the list ran on. We even spent some stormy weeks off the coast of Cornwall, in the hope of gaining sea legs, broadly coincident with the second lunar landings.[10] Unhappily, our own little adventure was beset by gales and spindrift—we were barely able to hold down food or haul up sand. Even so, as the embarkation day for the cruise of *HMS Fawn* drew near, we scuttled excitedly down to Plymouth to learn about sampling of plankton at sea, and then to the Naval Dockyard at Devonport to meet with the Officers and crew, to visit our cabin quarters, and to plan our cruise.[11]

At long last, our day of departure arrived. On 26 March 1970, we slipped away from England, standing to attention in formal dress alongside the Bridge. The bosun's whistle—a spine tingling sound at any time—was piped to *HMS Ark Royal* and other senior ships as we passed. Just as the *Fawn* rounded the promontory of Plymouth Hoe, I spotted three of my ancestors gathered in the distance, craning to get a better view, and waving their last goodbyes, until we—and they—were reduced to mere specks on the horizon.[12] It was a stirring farewell. And as we creamed through the waves of the English Channel, the waters behind the ship were marbled with green, like serpent skin. The 'beautiful green' of Robert Hooke could here be followed out to sea.

All at sea

Our twin ships, *HMS Fawn* and *HMS Fox*, sailed in tandem across the Atlantic at a steady 20 knots or thereabouts, a journey that took us some two weeks. On the outward cruise, we made an unplanned visit to Sao Miguel in the Azores after a nasty accident with an underweigh sampler, which scalped the top off an able seaman's head. We were soon steaming away again, though, and heading towards the Sargasso Sea in the middle of the North Atlantic Ocean. As we did so, a school of dolphins joined us, racing and twirling in front of the bows. Little by little, the sky got lighter, the sun got brighter, and the sea got smoother. Hour by hour, the world gradually took on the feeling of an altogether sweeter place.

Strolling onto deck one morning, the waters seemed as smooth as paint. But it was the colours of this world that really caught my attention. The Sargasso Sea itself was a deep and mesmerizing cobalt blue. And every five minutes or so, our ship would cross a long line of bright orange seaweed, seemingly as wide and as long as the wake of our ship. For just a few seconds, we would plough through this vast tangled mat of *Sargassum* weed, only to reach yet another long and quiet stretch of blue on the other side. The weed sat listless, seemingly rafting nowhere. From space, these mats must have looked like pin-stripes. For days on end, we pushed through these rows of orange weed and wide blue sea.

On one occasion, we hauled a tow net behind the stern of *HMS Fawn*, and then clambered up into the wet lab above the fo'c's'le deck, to explore our catch beneath a lens. Here in the Sargasso Sea, most of our haul consisted of Brown algae.[13] But, during our cruise, they displayed a delightful range of hues, like the tints of a Turner seascape: greens, ambers, reds, and browns. Colour, colour everywhere. What was causing such a variety of hues?

Meet the Greens

One of the most verdant of seaweeds we met with, emerald green in colour, was *Enteromorpha* (Figure 6).[14] In a seaside rockpool, this charming little alga mimics a bed of prize-winning lettuce. But encountered in a dockyard late on a dark night, this alga proved to be a very slippery customer indeed. This greenest of weeds not only clings tenaciously to the seafloor but also to bollards, dockyard walls, and the hulls of ships, slipping around the world on its long journeys unseen. Even saucier is its secret sex life. It can grow and reproduce so fast under raised levels of phosphate along polluted coasts today that it chokes the will-to-live out of some marine ecosystems, sucking up oxygen, and forming vast rafts of vegetation. These rafts have to be dredged out of the water by ships on an almost industrial scale.[15] This is not merely a seaweed. It is a rampant weed of the sea.

Such monstrous behaviour seems rather contrary to its slender appearance. Each delicate little weed is built from a gossamer thin sheet of cells, wrapped around into a tube rather like the emptied intestines of a sea snake.[16] When exposed to sunlight, these tubes undergo very high levels

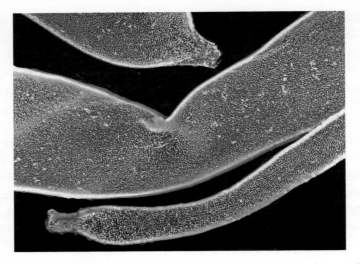

Figure 6. Rampant weed of the sea, called *Enteromorpha*. This Green alga is made up from cells that, down the microscope, can be seen to pepper the surface of each tubular, intestine-like frond. The largest frond here is just half a millimetre across but some can grow twenty times as large.

of photosynthesis and quickly fill with oxygen, which lifts them further towards the sunlight, a trick that must be helpful in the murky waters in which they often thrive. It is this intestinal appearance which provides the name *Enteromorpha intestinalis*, which literally means the 'gut-like' seaweed. These little giblets are constructed from myriads of tiny box-like cells, and each of these cells is infilled with handy gadgets such as nuclei and chloroplasts—the reproductive and photosynthetic structures found within each algal or plant cell. As it happens, both *Enteromorpha* and garden lettuce share the same kinds of chloroplasts, and the same kind of chlorophyll, too.[17] That is because both are actually rather distant cousins, albeit a billion times removed. Both have descended from a missing ancestor, a Green alga that was 'lost in action' long ago. Modern biologists are now able to tell us a great deal about this lost ancestor and its family tree. Molecular studies have shown that all land plants, ranging from bryophytes to *Bougainvillia*, and from liverworts to lettuce, can be found sitting alongside *Enteromorpha* on the same green branch of the Great Tree of Life.

Each member of this 'Green Party' looks appealingly verdant because of the colour of its chloroplasts—and hence of the wavelengths of light they absorb. Chlorophyll pigments may look green to our eyes, but green is exactly the colour of sunlight these plants do not wish to absorb, and so their chloroplasts throw this green light right back at us unwanted. Their green appearance is therefore a residue left behind by the main chore of the chloroplasts—that of absorbing, and then harnessing the wavelengths of red light.[18] So a red light actually means 'go!' to much of the living world. Of all the rays sent out by the sun's radiance—gamma rays, ultraviolet, and infrared—it is these red wavelengths of visible light that are best captured by algae and plants. Red light cannot penetrate down to greater depths of water because it becomes too rapidly absorbed.[19] No surprise, then, that all algal members of the Green Party thrive best near to land.

The Browns

The island of Greenland may have been named rather optimistically after chloroplasts locked within its sparse grassy meadows. And Elephant Island in the South Atlantic may have been named after its blubbery elephant seals. But *Sargassum* has the distinction of being the only creature to give its name to a large portion of the planet. The Sargasso Sea around Bermuda takes its name directly from the abundance of *Sargassum* weed, and not the other way about.[20] Today, this weed is clearly the most conspicuous living thing to be encountered here—other than the odd flying fish or dolphin, of course. At one time, it was even thought to cover the whole surface of the mid-Atlantic, making navigation across it all but impossible. Needless to say, that view was highly exaggerated, but we can sympathize with its drift. When great rafts of *Sargassum* weed were accompanied by feeble winds, a lack of automotive power, and a dearth of rainwater, this part of the ocean—including the infamous Bermuda

Triangle—was wisely avoided by sailors. *Sargassum* is still a weed to be regarded with distrust.

As we have seen, instead of being emerald green, this dreaded weed is amber coloured.[21] And instead of being like a flimsy snake's intestine, it is tough and rubbery, thrusting out bunches of wildly branching fronds into surrounding waters. When caught up within a tow net, its fronds can look a bit like the comedy hair of a clown—coarse and reddish. Looking more closely at this fright wig, we can see that some of its fronds display little amber beads. These are bubble-like bladders, called pneumatocysts, that help *Sargassum* to drift so widely across the ocean. So effective are these floats that one particular species, *Sargassum muticum*, is currently invading the world, hitching rides aboard fishing boats, or hiding in the bilge tanks of container ships. It has a prodigious rate of growth as well, clogging up marinas, fouling fishing lines, and growing to over twelve metres long. Another species called *Sargassum fluitans* is even more adept as a mariner. So adept, in fact, that it never attaches to the seafloor at all. It spends its whole life afloat, like an idle drifter (Figure 7).

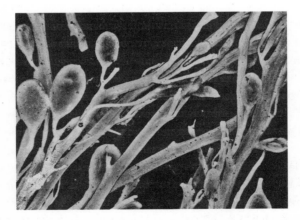

Figure 7. Fronds of the Brown alga *Sargassum* are held aloft in the water by the rounded bladders. The field of view in this microscope image is just a few millimetres across.

Sargassum sits on a highly distinctive branch of the Great Tree of life, nicknamed the Kelps, or the Browns.[22] Each of us unwittingly encounters the Browns everyday, but hidden from view—as alginate in our ice cream, and in our soap and toothpaste. Unlike the Greens, though, the cells of these Browns are never seen to live out their lives as individuals. Instead, their box-like cells are invariably clustered together into large colonies. Indeed, their colonies can comprise some of the largest organisms to be found alive on the planet today—rivals to the giant redwoods of California. Some Kelps, for example, are known to grow at the rate of half a metre a day, while the giant alga *Macrocystis* can reach lengths of eighty metres or more. So rapid is their growth, indeed, that the Browns may yet provide an invaluable source of fuel for the world. They certainly provide a source of food enhancer—the monosodium glutamate of Asiatic cooking is derived from yet another Brown alga called *Saccharina*.

The hues of these ancient mariners arise from the colour of their chloroplasts—ranging from dark brown, through amber, to almost orange in colour. Such russet pigments are rather efficient at working with subdued light levels—at high latitudes or in deep water where blue wavelengths predominate.[23] No surprise, then, that members of the Brown family tend to be seafarers by inclination. They have eschewed a life on land and are largely to be found at sea. And they are most diverse at sea as well. These facts alone suggest that Browns themselves did not evolve on land. Or even in lakes on land. They first evolved in the sea.[24]

Blue Greens

As we have seen, Green *Enteromorpha* had to be negotiated with great care around the dockyards at Devonport. And *Sargassum*—our Brown seaweed—gave fierce battle to our propeller blades whilst crossing the Atlantic. But it was not until we reached the safety of the harbour at Antigua that we came face to face with a third kind of weed—the Blue Greens—in great abundance.

Antigua is a long extinct volcanic island that lies athwart the Lesser Antilles island arc. Blessed with freshwater and pasture, it has for long been the first port of call on lengthy transatlantic voyages. Facing the Caribbean Sea along the westward coast is a delightful series of sheltered sandy bays. At sunset, the volcanoes of Montserrat can be seen from this vantage point, emerging as pointy silhouettes along the horizon. Across the island on the eastward side of Antigua, lie those bright and breezy coastal resorts that face into the trade winds. These breezes bring surf thundering in across the reefs from the Atlantic to the east. They also cool the fevered brow, and best of all, they brush away mosquitoes.

But none of these blustery virtues of Antigua brought much pleasure to Captain Horatio Nelson. What he needed back in 1784 was shelter from shifting sand banks along the west coast, and rough trade winds from the east. As expected, he chose to shelter his Man-O'-War in a deep inlet on the south coast, a place since called Nelson's Dockyard.[25] Nowadays, this inlet is decorated with a frothy mixture of yachts, chandlers, and pop stars. But during our visit in 1970, the harbour was still a rather sleepy spot. Its shoreline was still encircled by swamps and lagoons. And those lagoons contained a rich carpet of Blue Greens, among the most intriguing of which is a little creature called *Oscillatoria*.[26]

The sight of this little twister down the microscope can be quite unforgettable. Their cells are about a tenth as large as those of Green algal cells, so that from 500 to 1000 could be placed side by side on the head of a pin. The dwarf size of these Blue Green cells is compensated for, however, by their arrangement into long chains, each thread about a tenth as wide as a filament of fine silk. Being so tiny, their surface area to volume ratio is immense, for one very good reason—so that their constant demands for fresh nutrients can be met effectively.

The cells of *Oscillatoria* are not motionless in the manner of the Greens and Browns we have seen. Instead, they wiggle about. Oscillation, it seems, helps them to move either towards or away from the sunlight they need to process more food. Curiously, this wiggling can be done without the benefit of muscles—it is a purely chemical process.

At the time of our cruise, Blue Greens such as *Oscillatoria* were still called Blue Green algae. It was only in the later 1970s that both microscopy and molecules finally confirmed they were not really algae at all. True enough, they contain chlorophyll pigments in the respectable manner of algae and oak trees. But the chlorophyll pigments of Blue Greens are not clothed within special membranes, like those of algae and oak trees, but go about their business naked. And Blue Greens had no use for a nucleus at all. Instead, their circular molecules of DNA sit freely within the cell. Blue Greens are not algae at all, but true bacteria—eubacteria.[27]

The importance of this change in status for Blue Greens can hardly be overstated. Cells are the common currency of life—we are all made of cells. But some cells are more equal than others, it seems. Certain cells like *Oscillatoria* lack key ingredients such as the nut-like nucleus discovered by Robert Brown. Indeed, all bacterial cells lack a true nucleus. Such primitive cells are commonly called *prokaryotes*, meaning something like 'before [the evolution of] the nut' in Greek. Other cells, however, are more sophisticated.[28] They carry most of their genetic material in a nucleus and house a suite of handy compartments—which are actually reaction chambers—such as the mitochondria and chloroplasts. As we have seen, such chambers are called organelles. And the complex cells that carry them are called eukaryotes, which means something like 'complete with a nut' in Greek. Such organelles have never been found inside prokaryote cells such as those of the Blue Greens. Only the eukaryotic cells of the kind found in seaweeds—as well as in all protozoans, plants, fungi, and animals—contain a membrane bound nucleus and other organelles (Figure 8).[29]

In praise of walls

This distinction—between the grubby world of bacteria and the elite world of eukaryotes—can seem deeply puzzling. How is it that some cells remain simple, while others have become so complex during the course of evolution?

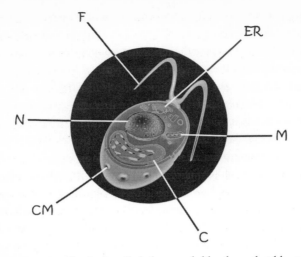

Figure 8. Diagram of a single-celled alga, much like those that bloom in the Atlantic, and cause the foam on beaches, here magnified many times. On it are shown some of the working parts of a complex eukaryote cell: (CM) a cell membrane provided with porthole-like channels; (N) a nucleus containing the recipes and regulations of DNA; (ER) an endoplasmic reticulum like a series of passageways, to localize transport and metabolic reactions; (F) two whip-like flagella for locomotion, attached to a basal body; (M) one of several mitochondria to serve as the site for the energy reactions involved in respiration and protein synthesis; (C) one of several chloroplasts, to capture photons from sunlight, and provide a source of free energy.

As we sailed across the Sargasso Sea, there was ample time for me to contemplate such puzzles and to pour through my pile of text books. Needless to say, I did no such thing. I was only 22, and I was much more interested in exploring the ship and its many oddities. Parts of the vessel had been made accessible so that I was able to wander about freely, climbing up into the Bridge, exploring the mess decks amidships, or poking about in the holds down below. It is only in retrospect that I can see how my jaunts through the ship were not so very different from exploring the inner workings of a eukaryote cell.

To begin with, the worlds of naval ships and biology each tend to speak their own special language. Some of these words are strangely shared—such as nuclear and fission. Some of them sound like ugly sisters—words like binnacle and barnacle. Quite a few are highly tedious and deserve to be avoided at all costs—pseudo words like charlie-delta-foxtrot and symplesiomorphy. But in general I have to admit that naval terms seem much more fun and win hands-down. The list of words that we heard as we crossed the Atlantic was very extensive indeed. Some naval terms were old norsemen's words—like Gear and Sky. Others, such as Banyan, were good old sailing ship words. A few, such as 'Andrew', have been torn from the lips of landlubbers by press gangs long ago. There were special names, we discovered, for all the naval ports—such as Pompy and Guz. There were words for every rank of seaman—such as Pusser, Buffer, Dusty, Swain, and Sparks. Collectively, the officers were all regarded as Droggies—hydrographers. As for our daily routines, we learned that we Dobied our laundry, ate our Pot Mess, drank our Mud, threw out our Gash, went to the Heads, and then slept—if we could ignore the rolling—in our Pits. There were, of course, quaint words for bodily functions that any schoolboy learns—though these were mainly applied to unsuccessful naval manoeuvres. There was even a pet name for the sea itself—the Oggin.

Thus armed with the naval lingo, we were now free to explore the elite world of HMS Fawn and its foibles; and the elite world of eukaryote cells and their quirks as well. Let us begin with the hull of our ship. It is hard to imagine any ship without its hull and its bulkheads. As our ship's surgeon, Paul Clarke, might have quipped, the integrity of any ship's bottom is vital to its good health. And if drowning is to be avoided, then the hull must be free of holes, and its bottom must be treated so as to discourage unwanted encrustations. Now, almost the same can also be said for living cells as well. Both cells and ships are tightly enclosed spaces, imprisoned by walls. It is no coincidence here that the words 'cell' and 'prison' are almost synonymous. More robustly too, we can admit there can be no life without walls. All of life is cellular. And all kinds of cells have 'walls'. That is why a virus cannot be regarded as a proper form of life—in part because it lacks that defining feature of life—the wall-like structure called the cell membrane.[30]

Cells are always enclosed by some kind of wall. So what is it about *walls* that matters so much to life? One answer, of course, is that walls allow for a better control of conditions, and hence for better concentrations of activity. This is true both within a ship and within a cell as well. In a ship, for example, there is a clear internal need for oxygen, for a stable temperature that is held somewhere between that of freezing and boiling, and for an absence of toxins. There is likewise a need for bilge pumps, to regulate the flow of liquids from the outside inwards. Too much seawater, and you have a Titanic on your hands. Too much alcohol—or maybe too little—and you have a mutiny on your hands. Compartments make for better regulation. So regulations are therefore needed onboard a ship, and equivalent regulations can likewise be found inside each cell.[31] Within the latter, for example, these regulations require the strict control of its oxygen supply—keeping it much lower than found in any ship—as well as the repelling of dangerous toxic elements such as lead and the hoarding of valuable ones such as phosphorus.

On HMS *Fawn*, strange to relate, a swift trade took place in molecules, especially after we entered a port such as Kingston in Jamaica. Regulations were particularly tight here with regard to two kinds of commerce across the gangway—guns and girls. The holding of guns was shunned, while the holding of girls was—and still is—regarded as entirely legal, unless the girls themselves object. So each night in Kingston, a matelot had to be placed on guard at the gangway. Needless to say, the main concern here was not so much the girls as the bacteria—gonorrhoea cells—that they might bring on board.[32]

Turning away from the foibles of our ship to the functions of our cells, a swift trade in molecules and elements can likewise be seen taking place across each cell membrane. There is nothing here that much resembles a porthole, a gangplank, or even a matelot on duty, of course. This commerce takes place instead through ingenious protein and phospholipid membranes that act as electrically charged pumps and filters. As this name implies, phospholipids are made up from oily materials called lipids, with added phosphate molecules.[33] On their own, such oily materials don't care to mix with watery liquids. They quickly form little cell-like droplets called mycelles—such a thing can be seen when shaking up a bottle of oil and vinegar when dressing a green salad. But molecules with added phosphate give an extra dimension to these lipids, making them 'amphipathic', a bit like amphibious frog skin. The inward facing pole of the phospholipid molecule is hydrophobic, meaning it will not dissolve in water but will associate together with other phospholipids spontaneously to form a chain. The outward facing pole is hydrophilic so that it has a greater affinity for water-soluble materials. Together they form what biologists call a lipid bilayer—the little secret behind the big success of the cell. Each bilayer forms a barrier that can be selectively permeable to water soluble substances of value to life.[34] Tiny pore-like channels in this membrane—just a few angstroms wide—act like portholes, for the transport of molecules into and outside of the cell, assisted by proteins. These behave like porters, carrying baggage.[35] In goes water. In go things like glucose, phosphorus, and potassium as well. Out go surplus things like carbon dioxide and sodium. And out stay big things too, such as 'polar molecules' with too much fondness for water.

For this transport to happen, the cell—like any ship—is therefore obliged to stay immersed in its watery medium. Water has remarkable properties as a solvent—so life is often called 'animated water'.[36] It will happily move through the selectively permeable cell membrane, from regions of lower to higher concentration, so as to try to equalize things on either side.[37] This movement of water is never quite enough on its own, though. The cell needs it to be tinctured with life-forming elements as well, of which there are about eighteen including, rather obviously, carbon, hydrogen, and oxygen.[38] Any cell removed from this nutritious briny medium is therefore likely to end up like a ship out of the Oggin: high and dry.

For curious reasons of history, not all cell membranes are quite the same. There are several different types of bacterial cell membrane. Some, for example, are seen to absorb colourful stains devised by a Danish doctor called C. J. Gram, which is why they are called Gram positive bacteria. My colleague Tom Cavalier Smith helpfully calls these Posibacteria.[39] Familiar examples tend to sound like a roll call of trouble makers, with names that include *Clostridium* and *Streptococcus pneumoniae*. Another kind, called *Agrobacter*, causes those knotted burrs on tree trunks that are much beloved by cabinet makers. But most bacteria are far from malign. Examples of this kind include the nitrogen fixers that grow alongside tree roots.[40]

Other kinds of bacterial cell membranes refuse to absorb the Gram stain, so these are called Gram negative bacteria, or Negibacteria. Notorious examples of the latter include the infamous Plague bacterium *Yersinia pestis*, as well as the equally infamous *Escherichia coli* and *Salmonella*.[41] But they also include beneficial forms such as oxygen-producing cyanobacteria like *Oscillatoria*.

But there is more. Not all bacteria have a lipid bilayer of the kind we have just seen. Only the true Bacteria—or Eubacteria—are found to possess this. Others, which have been placed by Carl Woese in a separate domain originally called the Archaebacteria—and later called the Archaea—have a rather different kind of wall, including forms with a lipid monolayer. These distinctive forms include the methane producers as well as the hot spring dwellers.[42]

Nor are plant and animal cell membranes quite the same. Those of animals tend to consist mainly of phospholipids, much as we have been talking about.[43] But the cells of plants and algae have a cell membrane that is largely reinforced with a glucose-based carbohydrate called cellulose, to form a structure called the 'cell wall'. It was this structure, indeed, that Robert Hooke first reported in those cells of cork down his microscope in 1665. And it was cell walls like this that we were spotting in our samples of *Enteromorpha* and *Sargassum* while crossing the Atlantic as well. In fact, you may be staring right now at the pulverised remains of plant cell walls—the pages of this book.

We should end by pointing to another way in which materials can be absorbed into the cell. This is the process of phagocytosis which means, quite literally, the eating of living or dead cells. Such predatory behaviour is made possible by infolding of the cell membrane to form an interior compartment called a vacuole. It is in this way that complex eukaryote cells like *Amoeba*, or indeed our own white blood cells, are able swallow up smaller bacterial cells, to store them for consumption at leisure.[44] As we shall see, this phenomenon of phagocytosis is not merely an ornament to our story. It is fundamental to understanding the ways in which our own and other complex cells have evolved in deep time.

The nuclear club

Trading not only takes place across the cell membrane, but also within the cell itself, and across the membranes that surround each organelle. We can therefore say that both cells and organelles have evolved into something that resembles a List of Ship's Regulations, to control this complex traffic in liquids and gases. Some of these regulations may be likened to strictures concerning bilge pumps or air conditioning. But beneath it all, comes a very serious concern—lapses in procedure can bring about an early demise. The inside of the cell needs to be kept electrically negative compared to the outside, and this is not easy to manage. There have to be strong checks and balances on the input of positive ions (such as calcium, sodium, potassium) and of negative ions (such as phosphorus). Too much calcium and the cell cannot access its sodium salts— its electrical conductors; or too much sodium and the cell interior becomes too salty and the activities of the cell break down. The list of regulations is large. It is tempting to envisage them printed in red ink and pinned up in the cellular gangway, complete with droll advice about abandoning ship, but they are, of course, written in the binary codes of nucleic acids strung along chromosomal DNA, the most ancient language known.

Interestingly for us, nuclear DNA does not just carry a list of regulations. It also carries the recipes needed for building the vital ingredients of the cell. Nor is the nucleus alone in carrying such lists and recipes.

Binary codes are also found inside two other organelles we shall meet, called the mitochondrion and the chloroplast. Taken together, these regulations and recipes help the whole cell work together as a single dynamic system.

Officers and organelles

As we crossed the wide Sargasso Sea, most of the nautical activities we have looked at so far were being carried out very efficiently by petty officers and ratings. So now is a good time to step forward to meet the officers—the big brass—themselves. And we must meet with a variety of organelles as well—for these are the officers of the nucleated cell. Highest ranking among the officers on board is of course the captain. On *HMS Fawn*, this was a dashing young Australian called Lou Davidson. He cared rather little for naval formality. Like Horatio Nelson, he liked to work hard and play hard. On one occasion, he even attempted his own version of the 'Nelson Touch', which is an old naval term for dare-devil leadership. Later in the cruise, while our ship was steaming along across the Great Bahama Bank, he oversaw the depth sounding while racing along beside the ship, towed behind a speed boat atop a pair of water skis. Clad in skimpy bathing trunks, he encircled *HMS Fawn* at dizzying speed, hallooing fresh orders through his megaphone as we steamed along. This not only amused the crew. It gave him a closer look at the patch reefs. Or so he claimed.

The real equivalent of the cell nucleus is not so much the captain, however, as his place of work—the ship's Bridge and Chartroom. The Bridge was nearly always a zone of intense concentration while we were away at sea. Officers such as Simon Richardson and Matt French, would stand on duty in tropical kit—white shirts and white shorts, white socks and white shoes, as though for a game of deck tennis. Needless to day, they were seldom allowed to play. Instead, these Navigation Officers were intensely consumed by the almost Masonic rituals of depth sounding and chart plotting. Back in 1970, those rituals also required arranging for our very own radio beacons to beam waves across the Caribbean, providing for a regular three-point fix—a Decca Hifix—on our position. There was, as well, an echo sounder to read off the water depths beneath our ship. And regular sample stations were needed to collect and record the nature of the sea floor itself—which is where we scientists came in. Behind this hub of earnestness sat the Chartroom, which was like an island of calm, where officers and scientists could work with less distraction. That was the theory. But distraction was around us all the time in the form of our briny medium—the Caribbean Sea.

Much as every kind of ship has its own Bridge, so every kind of eukaryote cell has its own nucleus.[45] Some even boast two or more. That alone warned early cell biologists that no cell can survive without its nucleus. But they were not content to let things lie, so they started to poke about down the microscope. At first, they found that cells from which this organelle had been removed were quickly found to die. While those into which an 'alien' nucleus has been transplanted were seen to grow into new cell types of the intruder kind,[46] like creatures from a sci-fi movie. In short, cell biologists began to discover that the nucleus is needed for a cell to reproduce, to feed, to grow, to regenerate, and even to change its patterns of behaviour. That is because the nucleus keeps tight hold of the recipes—the codes of DNA—which in turn produce the RNA needed for the manufacture of body-building enzymes.[47] Without these, the cell will simply turn turtle and die.

Each nucleus, like the Chartroom of a ship, is bounded by an outer wall that helps to keep its valuable equivalents of charts and records suitably ship-shape. This sack-like structure, called the nuclear envelope, is finely porous, which allows water soluble molecules to both enter and leave the chamber under a strict regime of controls.[48] Like the Chartroom too, this envelope encloses structures that can be loosely compared with book shelves—these being the ribosomes and chromosomes, as well as things on those shelves that resemble log books and maps—those navigational aids we call genes. When the Chartroom of the cell gets into a mess, or the charts go missing, then those cells will start to run aground. A sad story of this kind came to light when scientists studied tadpoles of the clawed toad, *Xenopus*. About one quarter of all its tadpoles were found to die owing to the inheritance of a genetic flaw: they could build nuclei but without proper ribosomal concentrations. Such embryos failed to hatch properly,[49] showing that even creatures like toads need accurate charts to builds their skins, warts and all.

Passageways

The inside of our survey ship was divided into a large number of discrete chambers connected by passageways—going up, down, and lengthwise. Navigating these in a ship that was pitching and tossing at sea demanded the formation of new skills. Scampering down a ladder while facing forward, for example, required a ludicrously flat-footed motion, with the feet spayed outwards like a clown. Walking along a passageway without hitting the maps on the walls or tripping over bulkheads beneath the feet, required a kind of drunken roll. At times, it felt as though we were learning the tricks of the circus.

We also learned that ships and boats are not interchangeable terms. Boats are something small that you can row. Ships are bigger vessels that need extra power. So following our metaphor, we can say that a bacterial cell recalls the smallest kinds of rowing boat, like a coracle—without discrete compartments. But a larger eukaryote cell is much more like a ship, often provided with extra power. Because they are about ten to twenty times larger than bacterial cells, they need to maximize their efficiency by forming lots of little local compartments, where reactions can be focused. This is made possible by dividing parts of the cell into networks of passageways, called the 'endoplasmic reticulum' in latin. This structure seems to arise from the outer membrane of the nuclear envelope.[50] But it is so small and hard to spot that it went entirely unseen until the invention of the electron microscope by Max Knoll and Ernst Ruska in 1931.[51] Such passageways greatly help to increase the reactive surface area of the cell interior, with winding membranes that somewhat resemble the meanderings of a maze. Along its walls can sit dark granules called ribosomes which are rich in RNA, and these carry out the synthesis of building materials like proteins. Streaming of internal cell fluid—called cytoplasm—around these membranes then helps to move enzymes and building blocks around. In eukaryote cells, the bulk of the interior can therefore seem bewilderingly dynamic—almost like a battle ship in full action. The innards of bacterial cells can seem placid by comparison, more like a dinghy.

Not to be confused with the endoplasmic reticulum is another set of structures in the eukaryote cell, called the microtubules. These tiny tubes act as a kind of skeletal framework for the cell—a cytoskeleton—acting like tubular scaffolding in a shipyard, holding together its complex compartments, maintaining cell shape, and maximizing surface area. Microtubules can also help to conduct materials around the cell, such as when chromosomes move about during reproduction.[52] Yet again, nothing quite like these microtubules is found, or indeed needed, inside bacteria.

Engine Room

From the Bridge of HMS Fawn, the navigational officer's coded instructions were conveyed downwards to the engine room by wire or phone. Down below, the engines, and indeed all things mechanical, were largely under the control of Petty Officers. Their domain included technicalities such as radios, radars, sonars, propellers, fuel, food, and even the Rum Tot. Like an empire set within an empire, this was never an easy world for an outsider to explore. That was, in part, because we scientists dined with officers in the Ward Room, rather elitely served by Stewards—called beagles—whereas the Petty Officers lived in a different way and on a different level. They met, dined, and drank in their own mess room, keeping for the most part to themselves. And they worked, more often than not, in dark corners with clanking machines.

These distinct domains onboard ship have their equivalents in the eukaryote cell, as well. If the Navigation Officers and the Bridge can be said to equate to the nucleus, then the Petty Officers and their domains—of bilge rat and black-hand-gang—can best be compared with that organelle called the mitochondrion. This organelle was first noted in 1857 by Albert von Kölliker who described what he called granules in the cells of muscles. Other scientists of the time noticed these in other types of cells as well. In 1886, Richard Altman went further and suggested that these structures were the basic units of cellular activity. In 1898, Carl Benda introduced the term mitochondrion from the Greek words for thread (*mitos*) and granule (*chondros*). Each mitochondrion is sausage-shaped and quite large—about one micrometre (a thousandth of a millimetre) long. Tellingly for our story, each is also about the size and shape of a single rod-shaped bacterium, though they can vary greatly in size and shape. Several such mitochondria can happily coexist within a single cell. Indeed, in the cell of some giant amoebae there can be ten thousand or more.

Each of these sausage-shaped organelles is surrounded by a distinctive double membrane. The inner membrane is complexly folded, to form a massively increased surface area for chemical reactions, rather like the multiple plates lined up inside the cells of an electric car battery. It is inside these sausage-shaped engine rooms that a powerful kind of fuel—called adenosine triphosphate (ATP)—is activated in the presence of water and then converted into adenosine diphosphate (ADP), with the release of energy. Such phosphatic compounds contain considerable stores of phosphorus—a highly explosive element used by humans in bombs and warfare. It is therefore ironic to reflect that some of the major reactions of this ATP were unearthed by Jewish scientists escaping the horrors of war. Carl and Gerty Cori,[53] for example, had been working together in Vienna in 1920 before fleeing to the United States to continue their research into anaerobic respiration—the famous burst of energy experienced by marathon runners as they cross the pain barrier.[54] Another refugee was Hans Krebs, a doctor who fled to England from Hamburg in Germany after Hitler came to power in 1933. He famously decoded what is now known as the Krebs cycle—the long and slow aerobic process that produces as many as 38 ATP molecules from one molecule of glucose.[55] In this way, explosive reactions within the mitochondrion have been found to provide the energies needed for locomotion, reproduction, and growth. A cell may use thousands of ATP molecules per second when highly active. That explains why cells with big energy demands tend to require large or numerous mitochondria. The human sperm cell, for example has a monstrous mitochondrion twisted around its tail, turning it into a self-propelled torpedo for urgent information exchange.

But there is a further curious twist to the story here. Not only does the mitochondrion come provided with its own membrane. It also houses its own DNA and its own ribosomal RNA, and these contribute towards its own growth and replication. Nor is this mitochondrial DNA and RNA much like that found in the rest of the surrounding cell. Instead, it shares properties with certain kinds of aerobic bacteria, including that of DNA arranged as a double helix of DNA within a circular molecule. This oddity provides an important clue for our story.

Sex and the single cell

Energy supplied by diesel fuel in the engine room of HMS Fawn was quickly conveyed downwards and backwards to the propellers at the stern—the main means of propulsion. Healthy rotation of propellers was considered rather vital to our survival at sea. Imagine the concern, then, when we ran into great rafts of Sargassum weed at night, and then heard the propellers munching grumpily through all those Brown algal cells.

Most eukaryotes have a comparable means of propulsion at some stage in their life cycle. It may take the form of numerous tiny hairs called cilia that beat like a Mexican Wave across the surface of the cell, as with a crowd at a football or baseball match (Plates 4a and 4b).[56] Or it may take the form of a flagellum, which spins around like a whip, to propel the cell through its watery medium. Sperm cells—which act as both explorers and messengers during sexual reproduction—are excellent examples of cells provided with such a flagellum. Nearly every kind of eukaryote has sex, so nearly every kind of eukaryote has flagellated cells at some stage in its life cycle.

But it turns out that not all sperm cells are quite alike. Indeed, eukaryotes can be usefully divided on the nature of sperm cells released during their life cycles. Some of these messenger cells have two tails—so they are called Bikonts, while others have only one—called Unikonts.[57] Curious to report, then, it is not our human superheroes that boast the machismo of two-tailed sperm. Male animals, and our cousins the true fungi, are only endowed with a rather whimpish, one-tailed sperm. Indeed, it seems shocking to confess that the burly, two-tail messenger is only found in algae and 'plants', but with one very intriguing exception— a group of highly successful single-celled blobs of jelly. These little blobs are the same organisms that build the tiny wreathed shells spotted by Robert Hooke beneath his microscope in muds dredged from the Thames estuary—Foraminifera, or forams for short. For 150 years, such forams had been regarded as close to the ancestors of animals like ourselves. They now stand exposed, however, not as the cousins of animals but as the cousins of plants. Indeed, they could be strangely close to the ancestors of plants themselves. That is because sightings of their two-tailed sperm, and of their molecular chemistry too, have recently given away their little secret.[58]

Sun Deck

Many of the features recounted so far may be found on most seagoing ships, and in most eukaryote cells too. But we now come to an even more exclusive world. The upper class domain of the top deck. As I discovered recently, the topmost decks of modern great cruise ships such as the *Queen Mary II* have been handed over to hedonism—a myriad of sun beds set out beneath a vast glass dome. It is here that white skinned people come to synthesize Vitamin D by exposing their pale skins to sunlight. Back in the drabness of the Ice Age, such an act of 'photo-synthesis' was no doubt welcome. Translucent skins helped to reduce cases of crippling bone disease such as rickets. Back on *HMS Fawn*, there was of course no such solarium, and no clear case of rickets, either. Indeed, there was no real need for photo-synthesis at all. Yet an area was still set aside where officers could sunbathe in privacy, set high above the Bridge.

This area of our ship was more properly called the Flying Deck. While it more closely resembled a boxing ring set high above the sea, it was nicknamed 'the Doctor's Upper Surgery'—because the Ship's Surgeon was not always treating damaged limbs of one kind or another, or fighting cellular invasions. There were times when his business was slack and he could climb aloft to sunbathe for an hour or two. Unhappily for him, though, we scientists often got there first, to lay out our specimens. The doctor was therefore obliged to sunbathe among a whiffy array of decaying corals and sun-dried conch.

In the sun lounge of the eukaryote cell, photosynthesis, with the release of gaseous oxygen, is likewise found, though here it is a vital activity. But we also need to admit that oxygenic photosynthesis can only be found in the poshest kinds of cell, provided with those green, brown, or red organelles called chloroplasts—the equivalents of solar panels.[59]

Chloroplasts are among the largest of the organelles to be found in eukaryote cells, being up to 10 micrometres across. They owe their colours—red, green, and brown—to different mixtures of various kinds of chlorophyll. In a single celled alga, there may be only one or two chloroplasts per cell. In other algae and higher plants, there may be hundreds in each cell. But their structure remains more or less the same—a double-walled membrane enclosing an inner system consisting of stacks of thin, flat, chlorophyll-bearing plates, called thylakoids. It is on and around these plates that the light and the dark reactions of photosynthesis take place. These light reactions require the absence of oxygen—they are anaerobic. And they require the photons of sunlight too. Absorption of these photons by chlorophyll on the thylakoids then results in the production of ATP from ADP, much as in the mitochondrion, but the reaction in the chloroplast also results in the release of oxygen to the atmosphere. Indeed, this is the very reaction on which we rely in order to draw breath.[60] The dark reactions, on the other hand, do not require sunlight and so can take place at night, in the spaces between the thylakoids. Energy stored in ATP is here used to convert carbon dioxide into carbohydrates—via the so-called Calvin cycle.[61] It is from this reaction that we, and the animals on which we prey, obtain the organic molecules we need for survival.

But there is a second curious story here. Not only does the chloroplast come provided with its own membrane. It also houses its own DNA and its own ribosomal RNA, and these contribute towards its own growth and replication. Nor is the DNA and RNA of chloroplasts much like that found in the rest of the surrounding cell. Instead, it shares properties with certain kinds of aerobic bacteria, including DNA arranged as a double helix of DNA within a circular molecule.

If you feel a sense of *déja vu* here, that is quite correct. For chloroplasts share these oddities with the mitochondria we have just met. Hence plant and animal mitochondria resemble a group of Gram negative bacteria called the alpha Proteobacteria, while plant and algal chloroplasts resemble other bacteria in that group—called the cyanobacteria.[62]

But there is a difference from mitochondria that we need to note here as well. Only algae and plants own their own chloroplasts. Others creatures that we shall meet, such as corals and forams, have had to borrow them if they wish to utilize solar energy. Seen in this 'organellist' way, it is not humans or even animals that mark the culmination of evolution. Plants occupy the very pinnacle of the 'organellist' tree of life. They sit high up on the Flying Deck of evolution because they sport more organelles than we do. Even on this upper deck of life, though, there is a kind of pecking order—not all chloroplasts have equal status. As we shall see, some are more equal than others.

Senility and celebrity

But hang on a minute. When comparing a ship with our cells, we are surely overlooking a very simple objection. On *HMS Fawn*, the captain had come from Australia. One navigating officer had hailed from South

Africa. The ratings came from all over the place, but especially from the Celtic coasts of Scotland and Wales, and the ship had been built in the shipyards of Lowestoft in England. In other words, the ship's crew did not have a common origin at all. It was like a composite entity with a nucleus from here, and a mitochondrion from there, and so on. Surely a eukaryote cell cannot have such a complex heritage, can it?

Take your very own cells for example. Your nucleus has been handed down from both parents. But your mitochondria come from your mother alone. And where are your chloroplasts? Have you lost them? How could that have happened?

Science, remember, is a unique system for the measurement of doubt. It throws up hypothetical answers and then tests them. And science is also what a democracy does—so at least two competing answers will be needed, to test for the fitness of their explanatory power. So let us look at two extreme solutions to this conundrum. Both sound a bit bizarre.

1. The earliest cells had all of these various organelles. Later forms have progressively lost them.
2. The earliest cells had none of these organelles. Later forms have progressively gained them.

The first of these statements—which we may call the Senility Route—simply pushes our dilemma backwards to the origins of life itself. It implies, at its most extreme, that we are descended from the ancestors of plants. And that bacteria sit at the end of the line, impoverished by an endless loss of faculties. These losses are acts of carelessness, as with a doddery old aunt who can never find where she left her spectacles. All this would be plausible, but for two salient facts. None of this story is supported by the fossil record—animals and plants were among the last things to emerge in the fossil record, while bacteria were the very first to appear. Nor is the Senility story line supported by molecular evidence. Animals and plants branch close to each other near the crown of the tree of life, to which the bacteria seemingly form a huge and diverse stem, as shown by Carl Woese and others.[63]

The second of these statements—the Celebrity Route—takes us in the opposite direction. It implies that novelties were added progressively to the cell through the course of billions of years of evolution. Life started down in the dumps, but built itself up gradually until it acquired all its accoutrements, like a celebrity, complete with mansion, sun lounge, and poolside bar.

But this hypothesis does not, of itself, tell us where those novelties came from. It is all very well saying 'and on the third day, the cell owned chloroplasts', but this does not tell us where or how the chloroplasts—the Greens, the Reds, and the Browns—themselves arose, nor why there so many different types, nor why they have their own DNA. Are they secretly planning a mutiny? And could chloroplasts really be among the most advanced things ever to have evolved on our planet? That is one of the questions we shall need to explore.

Blue Water

Seaweed was just one among many forms of life that we could spy from the upper decks while crossing the Sargasso Sea. From time to time, the surface of the sea would fizz with flying fish, skimming in straight lines across the water, cushioned by warm and rising air until they dipped and then were gone from sight. Every now and then, as we have seen, one of these little torpedoes would make a slight misjudgement and arrive unexpectedly upon our quarter deck. Those of us who had been gazing at this Sargasso scene enchanted, would suddenly scramble down the gangway to examine the crazy creature, and then fling it back to sea.

At first, it was tempting to see the Sargasso Sea in purely poetic terms—as a gentle lake within a fearsome sea. Over the following months, however, we learnt to see it in more ecological terms—as a Blue Water ecosystem almost beyond compare.[64] It was to be another two weeks, though, before we had the chance to examine these blue waters and their coral cathedrals more closely for ourselves. And, that would entail an encounter with a disappearing island.

THE PHANTOM ISLAND

Paradise lost

There is an island in the middle of the Caribbean Sea that no one living has ever seen. This concealed island—now called Pedro Bank—can be spotted on a sailor's chart, sitting well to the south of Jamaica.[1] Big palm leaves rustle in the afternoon thunderstorms of Jamaica and lush fruits glint in the following rains, while Pedro Bank—a large coral bank, almost as flat as a snooker table—sits beneath empty skies, set far out to sea. Seemingly, Jamaica has it all and Pedro Bank has almost nothing. A lost world, it sits brooding, just beneath the surface of the Caribbean, lying in wait for the next passing ship.

It was not always like this. Ten thousand years ago, Pedro was a twinkling jewel in the coronet of the Caribbean. Its romantic and isolated setting was then heightened by a curtain of steep sea cliffs that encircled the great island on all sides. But unlike the cliffs around Conan Doyle's Lost

59

World, or around the fabled Matto Grosso, there was not just one wall of great cliffs here. There were many, arranged both around and above the shoreline in something that must have looked from the air like a fortified compound, over a hundred miles long. During the rainy season, streams will have cascaded out of caverns to fall down this great staircase of cliffs, to flow across the bone white beaches and out towards the reefs beyond.

Thus embattled by cliffs, entry into Pedro must have required special skills. A scramble into the interior of this great island, some 130 miles long and 50 miles wide, would be met with shady caverns, sheltered hollows, and deep ravines, making any passage through the jungle here both slow and hazardous. Something like this can still be found in the deep interior of Jamaica today—called the Cockpit Country. That was a place where few people dared to venture, save for the desperate and the brave. In the Cockpit Country of old Pedro, though, there were never any human beings to take on such a dare. Hummingbirds fluttered unobserved around the cactus blossoms. Parrots squawked unheeded across groves of palmetto. And frogs and crickets croaked and chirped at sunset, to the annoyance of nobody at all.

This wilderness of Pedro sounds like a dream, but it was real enough. Indeed, a similar wilderness may return to the banks again, in the millennia far ahead. But no one can hope to bathe in its fabulous waterfalls or sup on its sugar apples just now, because the cliffs and grottoes around the island have been the victims of a huge shift in global climate.[2] These cliffs were doomed to slide beneath the sea as soon as the ice sheets of the last Ice Age began to melt, a mere ten thousand years ago. A vast tropical island, with all its living creatures and complex ecosystems, was thereby drowned in a geological trice. A paradise was lost.

Running the lines

Yet the land of old Pedro did not entirely disappear. Even in the modern world, there are still a few individuals who can revisit such localities, places remote in both space and time. They are called palaeontologists and they, or rather we, attempt to conjure up long extinct creatures. It was as a sorcerer's apprentice that I therefore grabbed the chance to reconstruct, for the first time, this long lost island of Pedro and to conjure up its dragons. And it happened like this.

It was a warm and windy night in April 1970. A few hours before leaving port, the crew of our ship HMS Fawn and her sister ship HMS Fox, had bid farewell to the bars and bordellos of Kingston, Jamaica. We then slipped past the ruins of Henry Morgan's pirate capital of Port Royal, and sailed out to sea.[3] Out into the wide ocean we sailed, where huge and rusty vessels would steam past us without warning during the night, conveying cargoes from Moscow to Cuba.

Our mission out here was to make a chart of the ocean bottom, just to the south of Jamaica, across the infamous Pedro Bank. To achieve this, we knew we would need to run the lines. And by that, we meant steaming along a pre-calculated line for hour upon hour until, at the sound of a bell, the ship would mercifully turn about and furrow back through the waves a mere four hundred yards to the west. We ploughed a wake up to fifty miles long, swinging along the ocean like the arc of a great pendulum, back and forth, again and again, for month after month.

Up on the Bridge on that night, the nautical machinery—radar and echo sounder, wind sock and navigator—were ticking away merrily in a corner. There was an air of great concentration. The Officer of the Watch was leaning over the chart table, with pencil and compass, and I was sitting before the two depth recorders, watching their revolving needles and calling out the water depths at three minute intervals. But there was also a niggling sense of anxiety. Our job was to map out several life-threatening hazards to shipping. For example, there were reputedly some patches of seafloor so devious that they kept sneaking about during the tropical storms which tend to hit this central part of the Caribbean every year, often with marked severity. These patches were seldom to be found in the same place from one year to the next. And then there were the coral reefs, hundreds of miles of them, just waiting to shiver the timbers of any passing ship like ours, and snack upon sailors.

But most haunting among the hazards we faced was the fear of hidden ship wrecks. Without modern satellite imaging to locate them, none of us back then knew quite where they were since their captains and crew had left their boats behind in a rush during sudden squalls. Hidden reefs and sudden tempests had always made this a perilous place to sail. Our job was to spot the shipwrecks before they spotted us and spoiled the jib of our ship, already damaged the year before on rocks off Scotland. The risk of hitting a shipwreck, or of becoming a shipwreck ourselves, focused our minds most wonderfully at sea.

We had no choice but to grapple with the nautically dangerous Pedro Bank and to somehow stay afloat. We were to map out the contours of this long lost island in detail for the first time. It took us five months at sea.

Atlantis found

My eyes were glued to the plot of the Echo Sounder as we traversed the south Jamaican shelf. For hour after hour, the plot had traced out a line of seafloor that lay deep below us. The flickering pen was hitting the paper with a slow and regular thud ... thud ... thud. But then, all of a sudden, the pen of the Echo Sounder began to whir into life, revolving frantically and picking out the profile of an underwater cliff line. This long lost cliff started to rise so rapidly towards us that I began to feel anxious. In no time at all, the water depths beneath our ship had changed from depths that could swallow up the Blue Mountains, to barely enough to cover a Jamaican Bus.

What we saw next, though, caused us to gulp with surprise. The flanks of old Pedro did not rise up towards us in one broad sweep, as I had rather foolishly expected. They rose in steps, rather like a giant staircase. I was fascinated to learn that Darwin himself had suspected such 'ledges' around coral islands.[4] But it was stunning to see it for myself on an echo sounder, for the first time. There, beneath us, was a succession of ancient shorelines, with cliffs and caves to landward, and lagoons and reefs to seaward, hidden deep beneath the surface of the sea. First, I saw a ledge about the width of a tennis court some hundreds of metres down, then another just 60 metres

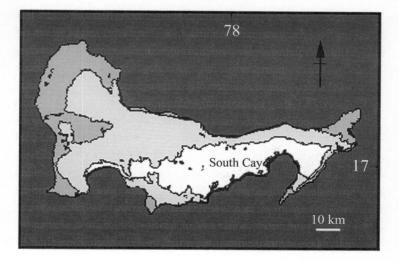

Figure 9. The Phantom Island, called Pedro Bank, now lies beneath the waves in the middle of the Caribbean Sea. This chart shows the depth contours drawn at ten metre intervals, as revealed by our survey aboard *HMS Fawn* and *HMS Fox* in 1970. To the north lies the Cayman Trench and Jamaica, while the tiny island of South Cay lies along the southeast perimeter of the bank, overlooking a large arcuate canyon system. Submerged coral reef terraces, which surround the margins, are too tightly packed to show clearly on this small chart.

down, and then another not far beneath us (Figure 9). Half a dozen such ancient shorelines were arranged in this way, some of them now so far beneath our ship that the sunlight could never hope to reach them.

As month followed month, we proceeded to map out the reefs and shorelines of this long lost world. Pedro Bank proved to be like a vast, lop-sided atoll with very patchy coral.[5] But reading the deep sea traces from the Echo Sounder always required great concentration. The seabed beneath the ship would plot as flat and monotonous for hours. And then a sharp pinnacle of coral would sprout up towards us unannounced, or the bank would fall away in steps towards the ocean floor. During the day, an able seaman would therefore keep a lookout on the Bridge for approaching shoals with the aid of a pair of polaroid sunglasses. To minimize the dangers of becoming a shipwreck ourselves, we decided to map the reefs and banks by day and to sound out the abyss of the deep sea floor by night. We were learning to become midnight explorers.

The Shipek Team

'Shipek Team close up!' came the order on the following morning, barked over the ship's tannoy. Dashing up to the deck, we could feel HMS Fawn starting to sweep round in a great arc, leaving a circle of smooth, clear, and deeply azure water within. At last, we were about to sample a Blue Water ecosystem.

As the wake of the ship subsided, we—Peter Dolan and I—attached a metal bucket to a great hunk of sprung steel, that infamously dangerous sampling device called a Shipek Grab. With a large iron bar, we anxiously primed open its jaws and set its catch plate—like a monstrous mouse trap—and then swung it clear of the deck and over the ocean. At this point, we had to stay steady or else the grab might swoop down upon us and snap shut on our limbs with bone crushing force.

The steel jaws of this grab device usually hit the seafloor, not far below us, in a matter of seconds. And as soon as it surfaced again through the surf, the Officer of the Watch would give the orders for the ship to 'steam on'. That was bad news for us, since it meant that we had to wrestle with the jaws of the grab upon the deck of a moving ship. Despite, or perhaps because of, the obvious dangers, a large crowd liked to gather round to watch the drama, and to help us examine our haul (Plate 6a).

Sometimes the bucket of the grab would contain a fist-like cluster of black sponges, looking like a bunch of 'deadman's fingers'. At other times, a haul of purple sponges came provided with neat little holes like the Pipes of Pan. More rarely, though, the grab contained a great tine of orange Elk Horn coral, *Acropora palmata* (Figure 10). This kind of coral is built from big blades of a mineral called aragonite. And it was this kind of coral that likely ripped out the bottom of Captain Cook's ship HMS *Endeavour* when he was trying to cross the Great Barrier Reef back in

Figure 10. A colony of Elk Horn coral, *Acropora palmata*, which helps construct reef crests around the Pedro Bank and elsewhere in the Caribbean. Its blades are typically orange or brown, owing to the presence of myriads of symbiotic dino-flagellate cells. These help the coral polyps to thrive in a nutrient-poor Blue Water ecosystem. © Franklin O'Donnell.

1770.[6] As we shall shortly see, Elk Horn coral has sent many a ship to the bottom on the Pedro Bank as well. Although this fearsome coral looks like a tree, and is covered in lots of tiny flower-like polyps, Elk Horn is truly an animal. In fact, it is a distant relative of the sea anemone, and of the jellyfish—a group collectively known as the cnidarians, from the Greek word *cnidos* meaning a stinging nettle. And as with jellyfish, its sting can be lethal to many forms of fish.

Each branch of Elk Horn is built up from a colony of polyps that number in their thousands. Neighbouring polyps are genetically identical, and produced by a process of vegetable-like budding. These clones behave like a single super-organism because they are connected by nerve networks and they share a common purpose: to capture animals, and to farm sunlight.

Every triffid-like polyp is beautifully adapted for catching little animals drifting about in the water column, such as the larvae of lobsters and small fish. It has myriads of tiny stinging cells called cnidocytes that harpoon tiny morcels of meat as they drift past in the current. Cnidocytes are one of the great marvels of cellular engineering in animals. Indeed they are among the most complex of all known cells, in terms of architecture. Each cnidocyte cell contains a tiny tube, coiled like a spring, which is poised to explode like a jack-in-the-box when its tip comes in contact with certain animal proteins. As soon as it is triggered, the coiled tube unleashes with spectacular speed, to turn itself inside out. Once inverted, the coiled tube brings into play its secret weapon—an outside sculpted with harpoon-like spikes so as to grip onto its prey. The hollow tube then acts with the force of a fireman's hose, pumping nerve toxins into its prey. Thus hooked and sedated, this prey—be it shrimp or snapper—is now doomed. That is because each coral polyp bears a ring of tentacles that wave around in the water, and wrap around any food item so as to pass it from rim to mouth. We almost expected to see the little slit-shaped mouth pucker with satisfaction, but these creatures never seem to smile. Once inside the gut, this space quickly fills with gastric juices that ferment and digest the prey, turning it into a nourishing broth. What happens next is highly insanitary. As with other kinds of cnidarian, such as anemones and jellyfish, the facilities for excretion in corals are shockingly primitive. Having no anus, all unwanted waste must therefore exit via the mouth. Few signs of sanitary, let alone intelligent design here.

A whirling dervish

The lack of an anus is not the only awkwardness faced by coral polyps. These corals are also coping with life in a Blue Water ecosystem. The water is blue because it is so clear. In other words, there is no river sediment to turn it brown and very little wind-blown dust to turn it yellow. And that means there is very little phosphorus or iron to feed the phytoplankton. It is this lack of phytoplankton—the tiny single-celled plant-like organisms that drift in the water column and feed by photosynthesis—that renders the waters so beguilingly blue here (Plate 5). Were they present in any numbers, then the waters would more likely be green or brown in colour, from the rampant growth of nutrient-loving algae. Corals would then lose out in the battle for space and light.

Coral polyps therefore live in something that can be compared to a marine desert. Almost no fertilizer, and almost no plankton. How is it, then, that they avoid starvation? The answer to this turns out to be very important for our story: they act a bit like pirates of the Caribbean. This astounding fact was being documented around the reefs of Jamaica by Thomas Goreau and his son—Thomas Jr—while we were conducting our survey of the Pedro Bank.[7] There followed decades of further dives and diversity studies, including those of a student of Tom Goreau Sr, Robert Trench, a Belizean who studied in Jamaica and then at Oxford. To cut to the heart of the matter, it was emerging that the cells of suitable phytoplankton species were captured and coerced to live inside the coral tissues as internal symbionts—or zooxanthellae as they are often called.[8] In this way, the entrapped cells of these symbiotic algae have been 'encouraged' to catch sunlight and manufacture carbohydrates—glucose, glycols, and amino acids—using photosynthesis. Employing cheap materials—carbon dioxide and water—they produce sugars in the presence of sunlight and chlorophylls. According to some recent claims, the coral receives up to 98 per cent of its energy requirements in the form of glucose obtained from these zooxanthellae.[9] It is a two-way trade. Algal symbionts, in return, receive some supplies of free fertilizer, as ammonia or other forms of nitrogen, sulphur, or phosphate, plus plenty of body-building carbon dioxide and a safe place to sit in the sun.[10] And the coral not only gets to dump some of its unwanted waste but is also able to gain valuable oxygen, plus extra energy for its expensive and ambitious project—the building of a coral reef. It is this fast-food relationship which keeps the batteries of a coral reef ecosystem fully charged and allows coral skeletons to grow with extra vigour towards the sunlight.

Such a relationship has its drawbacks. It means that reef-building corals are restricted to clear and well-lit tropical waters. Not such a bad deal, one might think. But there is a problem here. Like piracy, or slavery, or even like a marriage, this relationship can end up on the rocks. So before we look at the challenges, we therefore need to look more closely at the symbionts—the enslaved phytoplankton—themselves.

Studies of DNA show that each kind of coral tends to favour a particular strain or species of symbiotic cell called *Symbiodinium*.[11] This cell is a kind of dinoflagellate—meaning 'whirling whip'—of the type known to whirl like dervishes through the water column, capturing small prey as they go. Free-living dinoflagellates are remarkably versatile because they not only catch small prey but they also capture sunbeams for photosynthesis by means of chloroplasts provided with distinctive flame-coloured pigments.[12] Another distinctive feature of dinoflagellate cells living in the wild is their external battle armour—called thecae—arranged like the scutes of a Roman gladiator, complete with waistline and girdle.

Symbiodinium was first described by Hugo Freudenthal in 1962, when he found them living inside corals and in upside-down jellyfish as well. He also discovered that this enslaved symbiont performs a life cycle which is much like that of other, free-living dinoflagellates. That is to say, it had a blob-shaped vegetative stage—the one in which it lives inside the coral—as well as an armoured motile stage—in which it lives freely in the water column as phytoplankton, or nestling among the fronds of seaweeds. It seems that some coral polyps are able to capture these free living forms. A few are seeded accidentally by passing fish. Others, however, make sure to seed their coral eggs with suitable symbiont cells at the moment of ovulation. Once inside the cells of the gut of the coral, these highly favoured symbionts are not digested but encouraged. They can then multiply by a simple process of vegetative budding, like the algal slime seen growing on a garden path.

It has recently been shown, however, that dinoflagellates are in many ways quite unlike other algae. They are closer to a group of single-celled parasites called *Plasmodium* that bring about malaria.[13] Like those internal parasites, dinoflagellates clearly have an ability to live inside other cells for long periods. They also have their own distinctive machinery for photosynthesis. Most curiously of all, they have about one hundred times as much DNA as we humans have. Quite why they have accumulated so many genes and so much DNA, without the normal process of weeding, is bit of a mystery. Many appear to have acquired portions of their DNA from bacteria.[14]

In a Green Water ecosystem, where there are lots of nutrients, these dinoflagellates can whirl about freely in seawater. A few, such as *Gymnodinium*, can bloom so profusely that they turn the water as red as tomato soup. These 'red tides' bring about shellfish poisoning and fish mortality, especially during so-called El Niño years when tropical ocean currents become sluggish. Mercifully, such red tides are not yet known from coral reefs. Even so, the coral symbiont can itself manufacture toxins when in the free living stage of its life cycle. Coral polyps can therefore be said to have tamed these little demons of the water column, and kept them from doing much harm in the reef.

The world turned outside-in

Reef-building corals are therefore a bit like arable farmers—they manipulate the production of their symbionts. But while these internal symbionts can get along quite freely, meaning their part in this relationship is 'facultative', their hosts are utterly dependent upon them—or 'obligate'. Without their crops, corals would be completely unable to construct the reefs of the world.

Nor are corals the only animals at this game. Indeed, it can be said that reefs are hotbeds of symbiosis. Other cultivators include single-celled Foraminifera and radiolaria, multicellular sponges, flatworms, and molluscs such as sea slugs and the giant clam *Tridacna*.[15] Each of these involves a 'marriage' between an invertebrate animal and an alga, be it Green, Brown, Red, Golden Brown, or even Blue Green in colour. Close relatives of the Blue Green bacterium *Oscillatoria*, for example, have gained a passport to the innards of reef dwelling sponges, where they thrive as guests that pay for their stay with gifts of surplus food. One such symbiont is a blob-shaped cyanobacterial cell called *Synechococcus*. Some of the sponges that we found were orange in hue. Some of them were black. And just a few were a lurid sky blue in colour. The colour of these sponges is thought to reflect the different kinds of symbiont found within the tissue of the sponge, and all of them are thought to be bacterial.

But hang on a minute. Why do sponges need bacterial symbionts at all—don't they suck in bacteria constantly with the water through their pores? Well, it is true that a passing fish or lobster may stir up the seafloor nearby, and create a broth for sponges to filter, every now and then. But long periods of clear blue waters can make even a purple sponge feel peckish. Not only that, but the great warmth of tropical waters mean that the metabolic rates—especially of cold blooded animals—run at full throttle. That makes everything extremely hungry in a reef. Like corals, reef sponges will therefore supplement their gruelling diet by the farming of symbionts, and most especially, by the farming of bacterial symbionts.

On one occasion, I made the mistake of over-feeding a coral reef aquarium replete with handsome blue sponges. In no time at all, the coral rock became covered in a dark green slime. Beneath the microscope, I could see that this slime consisted of filaments—chains of those tiny Blue Greens called *Oscillatoria*. Wriggling constantly, these bacterial filaments moved about my doomed reef tank like a super-organism in search of light. Intriguingly, the filaments formed themselves together into a green cobweb that climbed up the reef rocks in a most uncanny way. This monstrous web was then observed to smother the sponges and then the sea anemones—even those that contained *Symbiodinium*. The poor reef animals obligingly died and decomposed in place, providing further nutrients for the cyanobacterial mat to imbibe. These bacteria ended up victorious, almost filling the tank.

Back among the reefs, the colour of a coral or a sponge therefore provides us with some hint as to the nature of the algae reclining within. Here, then, is a valuable clue towards that mystery we have called Green Waters and Blue Waters. It is as though the colour of the phytoplankton— reds, greens, and browns—has been sucked out of the seawater, leaving it beautifully blue and transparent. The colours of the phytoplankton have been transferred to the tissues of the corals, sponges, and other reef animals. That is why, in part, coral reefs can seem so festively coloured. It is as though the ecosystem has been turned inside out. Or more correctly, turned *outside-in*. As indeed it has.

A giant amoeba

Sponges and corals were not, however, the main rewards we were looking for within our Shipek Grab. Like children at the seaside, we wanted to gather up buckets of clean white sand. After months of sampling in this way, Pedro Bank was indeed found to be little more than a tropical marine desert. Its seafloor was a vast blanket of chalky white sand, rippled by frequent storms. That was a considerable surprise. But our next discovery was to provide a further telling clue.

One day, a gaggle of scientists and seamen had crowded round the bucket of our Shipek Grab, to see what treasures we had stolen from the bank below. And the floor of our bucket was satisfyingly covered with 'coins'; dozens of tiny gold 'coins'. I whisked the sample up to our Wet Lab above the Quarter Deck, my heart beating with anticipation.

This lab was an oak-panelled room loaded with sample jars and microscopes, secchi discs and sextants, atlases and log books. The room also smelled rather pleasingly of rum—presumably from the alcohol we used for pickling, though our amiable technician, John White, certainly seemed to spend a surprising amount of spare time up there as well. And here, away from the bright Caribbean sun, it was possible to examine those little 'coins' in detail, using a compound microscope developed from the example of Robert Hooke.

Scanning a tray of this sand under the microscope, I could easily see that these 'coins' were not made of metal. Instead, they were made from a chalky white material whose gold sheen was provided by what looked like a kind of yellow-green jelly—perhaps living tissue—hidden inside the disc shaped organism itself. Much more exciting than gold, I told myself.

A quick flick through several tomes on sea shells confirmed that we had stumbled upon a horde of giant protozoans—distant relatives of the amoeba. But these were not any old amoebae. They were grand dukes among the greatest of all amoeboid dynasties, and close relatives of the little Foraminifera that Robert Hooke had first seen and illustrated in 1665. Foraminifera are to geologists as fruit flies are to biologists. Let us make a brief digression to see why this might be.

Atomized atolls

The contribution made by tiny fossil Foraminifera to reefs and sediments of the oceans is generous, to say the least. This generosity first came to notice during soundings of the deep North Atlantic in 1856, in preparation for the laying of the telegraph cables in 1858 and again during the expeditions of *HMS Challenger* from 1873 to 1876.[16] The earliest attempts to drill through reefs in order to test Charles Darwin's theory were founded on Foraminifera as well. Professor William Sollas of Oxford took his drilling rig to Funafuti atoll and then bored the rocks—and no doubt all his crew—for months on end. But their efforts sadly came to nothing because the reef wasn't really made of solid rock at all. It was made of soft and soupy stuff: sand that was almost entirely made up of Foraminifera.[17]

The largesse of protozoans really came to prominence, though, during the first atomic tests at the end of World War II. From 1946, America was looking for a place where it could demonstrate the power of its newly developed atomic bombs. The military chose a chain of coral reef atolls in the Marshall Islands of the Pacific, including Bikini and Enewetak atolls. In the lead up to these tests, military engineers evacuated local fishermen and their families, and helped to sample the doomed reef shoals and lagoons. They also drilled deep down into the coral, not only to test the ground but also to test out Darwin's old coral reef hypothesis. It was from these tests that Dr Joseph A. Cushman of Sharon, Massachussetts was able to reveal the usefulness of Foraminifera for reconstructing the history of coral reefs in small samples of rock.[18] Like the eponymous Bikini swimsuit of 1946, his Foraminifera were found to be both small and revealing (Plate 3). And it was from these drill cores that his colleagues Ruth Todd and Doris Low were later able to extract Foraminifera to show how Enewetak atoll had started out life as a basaltic volcano some 50 million years ago. The debris of microfossils and coral skeletons amounted to a thumping 4100 feet—nearly a mile of reefal sediment.[19] They also showed, of course, that Darwin was right all along. His sun-loving reef corals and Cushman's sunbathing Foraminifera together built a rock that is seemingly afraid of the dark.

In truth, therefore, Foraminifera cannot be dismissed as junior members of the planetary board. They have been on the planet for a very long time—indeed, at least 540 million years or so—and the information stored in their shells is now proving invaluable for opening doors onto the deeper history of reef ecosystems and thence, as we shall see, for the whole history of cellular life itself.

South Cay

Every now and then, while undertaking this work, the path of our ship would pass a collection of small and uninhabited coral islands, strung out along the very southern edge of the Pedro Bank, in the middle of absolutely nowhere. These were true desert islands. Until then, I had not been able to see much of this strange world of sea shells and giant protozoans, save for specimens obtained from the mercurial adventures of our Shipek Grab. Anxious to take a closer look, I therefore put my name down to join a 'landing party' to one of these coral cays. But for weeks on end, the weather remained too rough—the Caribbean never stayed pacific—and the ship would buck about riotously in the afternoon trade winds.[20]

But at long last, a chance came to investigate a little desert island called South Cay—one of the smallest and most remote spots of land in the whole of the Caribbean. Mercifully, it was a calm and clear morning when the call came to 'lower the landing party'. The boat we needed on this occasion was basic—an inflatable provided with an outboard motor. Our main aim was to map out a hazard looking like a large brown rock sticking out of the shallows, just a few hundred feet south of our ship. It bore an air of malice.

As we approached, we could see it was actually an iron hulk, rearing some forty feet out of the water, with its back broken on the reef—an old coaster that had foundered in a storm. Ever more curious, we closed in with the shipwreck and climbed up the sides of its rust red hull to see if we could name the ship and date her demise. In a scene I shall never forget, we entered a bridge that was tilted at a high angle and rusting away, to find a mouldering log book still sitting on the table, complete with entries. She was the SS Maria. And she had been sitting there, through hell and high water, since foundering on the reefs during Hurricane Flora in 1963. It was like entering a haunted house, but set far out to sea.

Soon, we were speeding away again, to investigate the nearby patch of shifting cobbles called South Cay. My first ever desert island, it was little more than a pile of rubble, water worn and greying with age. Some of the cobbles were made from reef-building corals. But others were shells of a huge snail, called Strombus gigas, the queen conch, that had been swept up from their grazing grounds during tropical storms and dumped on this little hillock in the middle of the Caribbean. Bending down to scoop up some beach sand, I could see that the grains were the fragmental remains of sea creatures, corals, sea urchins, sponges, and snails, or of tiny natural sea pearls resembling the ooid grains first described by Robert Hooke in 1665.[21] Here, on South Cay, and nestling amongst conches and tiny pearls, we came across those giant Foraminiferan protozoans yet again. In fact, much of the sand hereabouts seemed to be built up from such skeletons, literally trillions of them.

Fool's gold

Many people tend to think of protozoans as rather inconspicuous. But in terms of diversity and in terms of bulk, it is single-celled organisms such as the Foraminifera that can truly astonish. Many seem to have been doing better in the recent past than they ever have done before, if the fos-

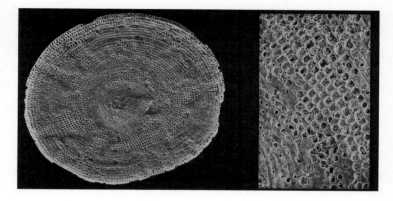

Figure 11. The chalky skeleton of a foram called *Archaias* has chambers arranged in a plane, like the grooves on an old vinyl record. Each ring-shaped chamber is subdivided into dozens of box-like chamberlets (best seen at right) for the efficient cultivation of symbiotic Green algae within its cytoplasm. It likes to attach to seaweed and sea grass blades in the backreef, where it may reach 10mm in diameter.

sil record is any guide. And best of all, these protozoans can be used as measuring sticks, to plot out much of the great history of life itself. It is time, then, for us to take a closer look at some of these curious callipers of evolution.

Those little 'coins' that we had hauled up to the surface with our Shipek Grab could be seen—under the microscope—to be tiny shells that were very thin and flat. Each was rather like a miniature 78 rpm phonograph record, though not much more than a few millimetres across (Plate 6b). And just like an old wax record, their shells were sculptured with ridges and grooves that spiralled outwards from the centre, ending up with a succession of complete rings. The whole structure was ivory white and very striking to the eye (Figure 11). But unlike a phonograph record, the shell was not solid. The spirals and circles marked out the positions of small internal spaces, called chamberlets, in which the living tissues of the giant protozoan have sat, storing up energy for its moment of release. After several years of feeding and growth, and on some kind of secret signal, its cellular mass would break up into pieces like a tiny jigsaw puzzle, each fragment swimming away from the shell to find a mate, or start a new life as a bachelor on the seafloor.

This extraordinary foram usually has a dome-like mound in its middle, making it into the shape that resembles a flying saucer. It goes by the latin name of *Archaias*. It is so abundant in the Caribbean Sea that its skeletons can form up to half of the bone white chalky sand we see today, spread across fabulous beaches from Belize to Barbados. And it is these skeletons that come tumbling out of our trunks and trainers as we enter the showers. More significantly, as we shall later see, these flying saucers can show us how the surface of the planet became carpeted with green.

This curious creature has turned itself, through the course of evolution, into a kind of biodome hotel in order to satisfy the needs of its algal symbionts. A single individual may contain many thousands of Green algae—the guests—each clamouring for a room with a view.[22] That is because these algal symbionts must have access to sunshine to manufacture the materials needed to pay the rent. This flying saucer-shaped protozoan therefore builds—from secretions of chalky matter—something that looks like a tiny hotel complex. The shell helps it to satisfy and sustain the demands of its algal guests. For comfort, their hotel is disc-shaped—to maximize the use of sunshine—and it can be up to a centimetre across. It is divided into hundreds of neat little box-like rooms—Japanese style—each provided with a nice glassy view of the sunlit sea, plus two handy doors for trade. Commerce is needed not only to bring essential ingredients into the sunlit guest rooms—such as nitrogen, phosphorus, and iron—but also to export the payment of rent back through the hotel and down to the treasury of the host. Needless to say, there are also corridors that connect the various suites of rooms, plus stolons—stairways—up and down between the levels. Every little wish seems to be catered for here. When the time comes to move on, to a new beach frontage perhaps—both the protozoan and its many guests swing into reproductive mood in tandem, like partners in a tango.[23]

Figure 12. The shell of this foram *Borelis* has chambers wrapped around an axis, like a rolled up carpet, as shown in the cross section to the right. It likes to knock about in current swept reef sands where it cultivates symbiotic diatoms within its hundreds of tubular chamberlets. This shell is about 1mm wide.

Football teams

Protozoans like this flying saucer called *Archaias* are well adapted to lounging around on flat surfaces, such as seaweed, sea grass, and lumps of dead coral, but they seemed to us a little too thin and fragile to cope with conditions near to the reef itself. To test this, John Scott and I took a dinghy out across the reefs, in order to take closer look.

The coral reefs hereabouts were for the most part sheltered by a great rim of Elk Horn coral, and the waters were a hypnotic blue colour. As we navigated away from the shore, though, we could see that the floor of the backreef lagoon became darkened with patches of dark green seagrass. It was both on and around the sea grass that *Archaias* often seemed to be living. Moving further away from the shore, the waters increased to a depth of about three metres and seemed largely bereft of vegetation, allowing the formation of a wide blanket of rippled sand. Being so close to big waves that came crashing across the coral reef, there was a strong current here, and many creatures—corals and crustaceans, sea grass, and seaweed—were evidently finding it hard to gain a foothold. This was the zone of the sand flat.

Life in the sand flat can be a bit hellish. It somewhat resembles life today on the dusty perimeter of a building site. The first problem is that wasted debris from this building site—the coral reef—comes sweeping across the sand flat, sometimes in ship loads. Worse still, seasonal tempests tear up the sandy seafloor, smothering many of its inhabitants. Rather few life forms other than Foraminiferan protozoans can flourish in the harsh conditions here. And even for them, life is little better than in a desert—sun, sandstorms, and near starvation.

With sand grains kicking around all the time, nature has therefore encouraged the evolution of a protozoan that looks as smooth and rounded as an American football (Figure 12). Called *Borelis*, it is a close relative of *Archaias*. But instead of winding its tiny chambers around a narrow axis, it wraps them around an elongated axis like a carpet, to make a football-shaped shell.[24] Like a football, such a roundness allows it to roll around the pitch without coming to harm.

Drama queens

Finding a passage through the crest of a reef, built from serrated blades of Elk Horn coral, was always a nerve-wracking business. As this zone was approached, we found that the waves tended to build higher and bouncier until they broke into a great dancing wall of surf. Such white water never seemed like a good place to linger because the water was exceedingly shallow here and the coral blades proved as sharp as bayonets. Great jumbles of jettisoned coral could also be seen sitting wedged together by the waves. A slight misjudgement, and our dinghy could have been thrust onto jagged rocks. We were keen to avoid the mournful mistakes of *SS Maria*.[25] Happily, there was a narrow channel through the reef that could be negotiated with care—the sea flowed inwards through this gap with great force.

Looking down into the water of the reef channel, we could spy the gothic spires of coral and gaudy pipes of sponge overhanging its steep walls. Surfaces here were carpeted with chalky crusts that were mottled with pink and white, like a strawberry yoghurt. These crusts were being secreted by another group of algae, called the Red algae. This important group—which Darwin called the 'nullipores', and today are called the 'corallines'—has a distinctive set of photosynthetic pigments.[26] Those that live in the reefs, such as *Lithophyllum*, often have calcified cells walls that look remarkably like the cork cells of Robert Hooke when viewed down the microscope. Their red pigments actually allow them to cope with living in quite deep waters or in low light levels.[27] But this means that Red algae can therefore cope with the loss of light that results from calcification. And that calcification then allows them to deal with a whole range of stress found in reef waters, such as voracious invertebrate grazing, high levels of wave action and damaging ultraviolet light. Indeed, their chalky crusts help to bind the whole reef fabric together. Corals provide the scaffolding of the reef. Red algae make the cement. Without them, the spires and domes of corals and clams would simply be nibbled to bits.

Looking down into the bottom of the channel, it was possible to see carpets of fresh white sand. It was here in the sand of the channels that the football protozoan called *Borelis* seemingly liked to knock about best. Sampling in such a place was a sticky business, though, because of very strong currents that swept through the channel. Like Darwin on the *Beagle*, we were therefore obliged to use a sticky plate sampler, this being a metal sheet covered in grease.[28]

Once we were seaward of the crest, we found that depths increased rather rapidly, and the corals dipped quickly out of sight. In no time, the water became too deep to dive, and all our sampling had to be done, either with that sticky plate sampler again, or with a hand-held set of metal jaws arranged at the end of a long line—a tediously unreliable device called a Van Veen grab. It proved tedious because it wouldn't collect sand, only a few well-washed coral sticks, and at times not even those.

With sample jars in hand, we collected scrapings of debris from these coral sticks in front of the reef. Back on board ship for the afternoon, I peered down the microscope at our catch. First to come into focus were the single cells of Golden Brown diatoms (Figure 13). At high powers of

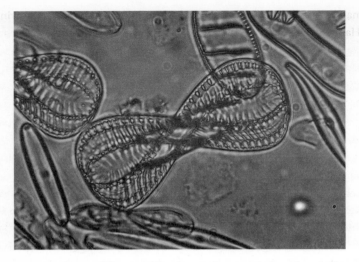

Figure 13. A cluster of fancy Golden Brown diatoms, each with a shell intricately constructed from glassy silica. The field of view is about 3mm across.

magnification, some of their golden filaments looked a bit like the drama queens of the cellular world. Their shells are moderately large (fewer than about 50 cells can be placed end-to-end in a millimetre) and they can appear a little overdressed, an impression heightened by their complex, almost fairy-like external sculpture—two valves of opaline silica bearing fussy little spines, all shot through with prissy little pores (Plate 4b). These valves—called frustules—can look even fancier than an Elizabethan ruff, with surfaces that are flounced, gingham'd, or gadrooned like taffeta lace.[29]

Staring at diatoms down a microscope on a long winter's evening was therefore—in the days before cinema—a favourite private pastime for many a Victorian gentleman. With the possible exception of the Crystal Palace—itself constructed like a giant pinnate diatom—these bugs may have been among the most complex glassy things that their admirers were ever able to set their eyes upon. Dozens of little frustules of this kind were sometimes painstakingly arranged by the English microscope slide makers—such as E. T. Norman or C. M. Topping—into messages saying 'Happy Christmas' or into immaculately choreographed circles and stars anticipating a chorus line. As cells go, however, diatom cells are not especially sexed-up. In fact, they reproduce asexually by dividing into two (binary fission) about once a day, and put aside the sex act except for once or twice a year. Although largely celibate and tiny, diatoms can still be rated among the big players of the living world dominating, as they do, so much of the carbon budget of our planet.

A further twist

It was in this rough zone of rubble that we occasionally encountered our third giant protozoan, called *Heterostegina*. This Foraminiferan can look like the twisting shell of a living *Nautilus*, or even the shell of a fossil ammonite. Both of the latter were cephalopod molluscs, of course, with squid-like bodies provided with a battery of tentacles and highly developed eyes. Hence, when this twister, *Heterostegina*, was first discovered back in 1826 by a Parisian geologist called Alcide d'Orbigny in a phial of Cuban beach sand, he thought it to be a kind of tiny cephalopod. Indeed, all those giant protozoans that we now call Foraminifera were once thought to be tiny mollusc shells, and relatives of the squids and ammonites. It is true enough that they both shared shells that were coiled and divided into little rooms called chambers, and that each room came provided with an opening called an aperture. That may explain why, when Denys de Montfort of Paris first reported those flying saucers and footballs in his book 'Conchyliologie' back in 1808, he felt safe in calling them molluscs.

But appearances can be deceptive, and fossil appearances are often so. Palaeontologists and zoologists have often found that similar looking organisms have evolved independently, from very different ancestors. One need only think here of living bats and similar looking extinct pterosaurs, or of living dolphins and extinct ichthyosaurs, or of living wolves and extinct thylacines. This list of look-alikes, called homeomorphs, is a very long one indeed.

The truth about Foraminifera first dawned in 1835 when Professor Felix Dujardin of France realized that they are not really very mollusc-like at all. First of all, they are often much smaller—down to 1/50th millimetre across—and they also come in a much greater range of shapes. In addition to that, they can make use of rather strange building materials—such as sponge spicules and mica flakes. But even more telling was the discovery that each individual foram consists of a single cell with an elastic outer wall. So elastic was the wall, in fact, that it commonly grew into root-like extensions called pseudopods. He therefore termed this group the Rhizopoda—the root-footed cells. It is this elastic outer wall, and its ability to swallow up other small creatures that makes these forams so crucial to our hidden history of the cell.

The twisting shell of *Heterostegina* is actually a bit of a pin-up among living protozoans. It enjoys this status in part because of pioneering researches undertaken by Rudolf Röttger of Germany in the 1970s. Collecting specimens from reefs of the Caribbean and Pacific, he spent many years watching this protozoan grow, feed, build its shell and then go through its many and rather confusing matrimonial arrangements. Röttger was also able to take videos of their flexible rubbery pseudopods, and their endless searches for food. But most importantly, he was able to show how this twister was able to survive without ingesting any external sources of food at all, especially in conditions like the Blue Water ecosystems of the Pedro Bank. It seemed able to flourish entirely by taking in other kinds of cell as symbionts. These cells were not dinoflagellates of the kind we have met living in corals, but cells of Golden Brown diatoms.[30] In other words, *Heterostegina* was not living alone. It was effectively farming diatoms, either digesting them or feeding on their photosynthetic products. Like reef-building corals, these protozoan cells have been taking in strange lodgers.

This was not the first time that protozoans had been caught making secret liaisons with photosynthetic unicells. Ever since the seventeenth century, people had marvelled at the antics of a single celled creature called *Paramecium* found thriving within a drop of pond water. Coated with tiny whips called cilia, this slipper-shaped protozoan would bustle about like a fussy farmer on market day.[31] But was it an animal or a plant? The way it bustled was very animated—so it was thought by some to be an animal. But it was often filled with green granules too—so others called it a plant. This dilemma was eventually resolved by Frenchman Felix le Dantec in 1892.[32] Mixing together some *Paramecium* with granules and others without, he found the latter quickly gained them from their neighbours. And there was more to report. When the lights were switched on, they all stayed green. But when they were switched off for days, his little beasts turned brown. These tests showed that their green lumps were not home-grown granules at all. Instead, they were Green algae that had been captured from surrounding waters and then encouraged to live as symbionts within the cell.[33] Le Dantec had stumbled on a world turned outside-in (Figure 14).

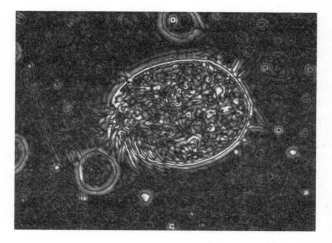

Figure 14. Protozoans, like this one related to *Paramecium*, have a mouth-like depression in the cell membrane for the intake of bacterial particles (at left), surrounded by a fringe of whisker-like cilia, for movement. They can also contain a scattering of single-celled Green algae which are farmed for glucose (light spots within the cell). This single cell, which was spotted on a frond of the seaweed *Enteromorpha*, is less than one tenth of a millimetre across.

Symbioholics

The word 'symbiosis' simply means the living together of differently named organisms.[34] Not all symbioses are benign: rats have fleas, and tigers get ticks. But nature has a way of encouraging any kind of interaction between organisms if they can turn a profit together. And most collaborations can and do turn a handsome profit, much as we can find in human society today. It is no surprise, then, that Darwinian evolution has encouraged some protozoans to compensate for their small size—and their metabolic limitations—by throwing in their lot with other small creatures provided with usefully different skills. A protozoan 'host' can capture bits of passing meat from time to time by trapping them with sticky pseudopods. This helps it to manufacture fertilizer in the form of phosphate and nitrogen compounds. And the symbionts can then capture sunlight and use this to manufacture sugars and other by-products, even when the supply of food is low. These by-products are available for trading with the host in return for any offer of free fertilizer in the form of phosphate, nitrate, and carbon dioxide. Both guest and host can be winners here as long as there is plenty of sunlight, which is usually the case in shallow tropical waters. Indeed, the pigments in these symbionts may act like a sunscreen, blocking out dangerous harmful solar rays. Together, such teamwork can prove remarkably efficient. Algal symbiosis can out-compete other creatures with lesser social skills.[35]

Not everything in the reef is symbiotic in this way, of course. Some creatures disdain it completely. Others—such as sea anemones and certain kinds of protozoan, have taken up the cultivation of symbionts as a kind of hobby. They find it helpful but they have not yet become dependent upon it.[36] Some only utilize symbiosis during summer months ('cyclic symbiosis'). Others only gain their symbionts when they are adults.[37] Yet many organisms in the reef show increasing patterns of dependence upon their symbionts. This dependency can reach the point where they can neither feed nor reproduce if their symbionts have either died or vanished. They have become completely reliant on their lodgers.

This gradation—from moderate to greater dependence—calls to mind varying levels of drug dependence in humans. Thus we can say that an ecosystem may contains some symbiont tasters, a category that includes various kinds of sea anemone; keen symbionts fanciers, among which are many soft corals; and the real symbiont users, who are downright hooked, including most reef-building corals, or those Foraminiferan protozoans we have met—the flying saucers, the footballs, and the twisters.

It is these desperate 'symbioholics' that have a great deal to teach us about the hidden history of the cell. They have created chimaeras—strange combinations of forms like the Sphinx of the Nile, half man, half beast; or like the Griffin, half eagle, half lion. But these chimaeras are stranger still. For here we have one organism sitting inside another. In the reefs we encounter the strange paradox of a *Sphinx Within*.

Out of their depth

It may be no surprise to learn that giant protozoans such as the flying saucer *Archais* have been found to farm single-celled algae whose photosynthetic pigments are best-suited to their preferred habitats. Some species therefore sit upon seaweed in brightly illuminated waters, tending to farm Green algal symbionts, while others can live at depths of 150 metres in the forereef, and prefer Golden Brown diatoms. Yet others at intermediate depths, harbour flame-coloured dinoflagellates. Interestingly, each host is not only highly devoted to its lodgers, but also secretes a shell ideally suited to keeping them happy. When this consortium is starved of light for weeks on end, the host will also find it difficult to keep growing. Protozoan landlords and their algal lodgers will commonly reproduce in tandem too, clinging together when their moment for reproduction comes along. This is clear evidence that the hosts have become symbioholics.

In this way, we find that Darwin's beloved coral reefs are full of deeply bonded relationships. Corals do it, sponges do it, even articulated clams do it. Sometimes it seems as though all the algae in seawater have been sucked inside animal tissues or inside the cells of giant protozoans—

leaving the seas themselves remarkably clear and blue, as we noted earlier, and contributing to the great riot of reds, greens, ambers, and blues that makes the reefs alluring.

Yet, does any of this race for maximum efficiency really make much sense? Why, you might ask, does not the coral or the protozoan simply release its waste products into the water, to encourage more plankton? And why doesn't the symbiont reject the advances of its host, or hang about in the water until a big storm stirs up the nutrients? No doubt these things will happen from time to time. But a fine balance has to be drawn in nature, as Darwin clearly saw. The outcome of this struggle for dominion of the reefs will simply depend upon which strategy—the collective action of symbiosis, or the freedom of separation—can generate the greater number of offspring with the better chances of survival. It doesn't have to be the best. It only has to be a little bit better than all the rest. As soon as that happens, then the game can be won—in the fullness of time.

In such a way, a tight bond may grow between the host and its symbiotic guests. It would be wrong, though, to liken this liaison to a life of eternal bliss. When looked at from a different perspective, photosymbiosis can look like a very dangerous liaison indeed. To find out how and why Blue Water ecosystems collapse, then, we had to wait until the next leg of our cruise aboard *HMS Fawn*. In July 1970, our ship had begun to steam southeast along the coast of South America, headed for a refit— both mechanical and mental. But we were also in for a sobering surprise.

CHAPTER 5

MOUTH OF THE DRAGON

Orinoco welcome

After several months at sea, nearly forty of our deck hands had sustained injuries from boat party adventures on the infamously choppy Pedro Bank. Dangerous shallows and lack of shelter during the hurricane season had therefore made it seem a far from idyllic place. Frequent squalls, big waves, and lots of white water made work both gruelling and dangerous here. By the end of July, we were therefore glad to leave this phantom island far behind.

Our new berth at Chaguaramas Bay in Trinidad therefore felt more like paradise. The skyline hereabouts was decorated with pointy blue mountains, the air was filled with parrots, and butterflies fluttered silently in the shadows. All around us, the sea seemed strangely calm. There was one niggling little problem, alas: Trinidad was in revolt. Drunken rioters were running through the streets at night, burning cars, smashing windows, and stealing goods. A stride into town saw us accosted by the hawkers of loot—wrist watches, TVs, and booze. We had chosen a rum moment to arrive.

A few days before this adventure, we had sailed around the northern coast of Venezuela. From the bridge, we surveyed a sharp line in the water ahead—blue and choppy on our side, and brown and smooth on the other. In no time at all, we were steaming across this mysterious marker, the boundary line between the waters of the Caribbean Sea and those of the Orinoco delta. We were entering the Bocas del Dragos—the Mouth of the Dragon.

While sailing through these waters, we quickly became entranced by the contrasts between life across this Caribbean–Orinoco boundary. On the Caribbean side, coral reefs abounded, in a typical Blue Water ecosystem.[1] But further south, around Trinidad, which sits astride the Bocas del Dragos, there were no such reefs. It is true that the island boasted wide beaches with coconut palms, but the sands here appeared yellowish in colour rather than bone white. And when viewed down the microscope, such sediment revealed little evidence of those giant protozoans with their green or golden symbionts and their endless obsessions with efficiency. There were no signs of the flying saucers, footballs, or twisters. Most of the foram shells hereabouts proved to be of a simpler, scruffier kind, like Robert Hooke's *Ammonia* from the Thames.

Not least of our puzzles, therefore, was that these waters around the island of Trinidad seemingly lacked conspicuous reefal growth—of either corals, forams, or fancy conch shells. That is not to say the seas around here were barren. Indeed, the waters truly teamed with life. But it was swimming or floating in the water column, or scratching about on the bottom. It was not decorously rooted to the seafloor in great fanfares of coral. If there were any corals to be seen hanging around here, they seemed to be of a solitary rather than colonial kind, and they were not much interested in photosymbionts.

'The entire absence of coral reefs in certain large areas within the tropical seas is a remarkable fact' wrote Charles Darwin.[2] There were none for him to discover, for example, around the Galapagos islands, which span the equator on the western side of South America. Darwin was inclined to explain this lack of corals and symbionts around tropical islands like Trinidad by suggesting the land hereabouts was rising out of the sea, thereby killing off all its corals. That might have seemed like a neat get-out clause for their absence around volcanic islands like the Galapagos. But it could not work for Trinidad because the island is not volcanic, and it has not been rising conspicuously out of the sea for some time.[3] Which means there had to be a better explanation. And indeed there was: we had sailed into a Green Water ecosystem.

Pet shop philosophers

The terms Blue Water and Green Water are, of course descriptions rather than explanations. They are observations of a *pattern*, rather than predictions about a *process* going on in the sea. To remind ourselves how thinking about pattern and process may work here, it may be helpful to recall the work of Charles Darwin and his theory of coral reefs. It took Darwin some five years or so—on and off the *Beagle*—to gather the information about corals around the islands of the world—from his own observations, as well as from books and private letters. As we have seen, he had noticed a *pattern*, ranging from volcanoes to atolls. The conception of a *process* that could explain this pattern—virgin volcanoes that ascend, to ancient volcanoes that descend—actually emerged before his collection of the data.

It has to be said that, in Darwin's day, it was a bit of a scandal to arrive at a hypothesis—a scientific idea—*before* all the data have finally come to hand. That is because one of the founding fathers of science—Sir Francis Bacon—had forewarned about the dangers of empty speculation, back in the early seventeenth century. The world before Bacon was infamous for its speculation. Much of it was fanciful or just downright silly. In place of foolish armchair philosophy, Bacon suggested that we take a gradual approach. He suggested that scientists should carefully assemble an empire of information before hazarding a hunch about the processes at play. This empirical method was considered the true path to respectability within the grand halls of English science. The opposite approach, that of an inspired hunch, or even an inspired dream, was considered a little too metaphysical, too Continental.

By 1830, however, it had dawned on a few Englishmen, like the astronomer John Herschel, that the way in which an explanation emerges in the mind does not really matter a jot.[4] It could emerge gradually, or it could arrive unexpectedly, from the unconscious and out of the blue, for reasons that Sigmund Freud would later explore, following a hunch of his own. All that really mattered in science, said Herschel, was that a good idea be allowed to emerge in one way or another, and that the idea then be properly tested.

Sometimes, big ideas will come before observations, as with Darwin's coral reef theory.[5] Sometimes they will emerge during the work itself, as with many student theses. And sometimes they arrive well after the event, as with Darwin's big idea about Galapagos finches. It is fair to admit, therefore, that it took many decades for scientists to appreciate the deeper meaning of coral reefs and Blue Water ecosystems. Some of this reality arose from further sampling of the seawaters by oceanographers.[6] Some of it emerged from laboratory experiments. But a most striking reality emerged from a very strange quarter indeed: from pet shops around the world.

During the final decade of the twentieth century, there was a rising demand for curious and exotic pets. Small boys were demanding spiders. And their larger siblings wanted snakes or snappers. A rise in scuba diving was likewise driving a demand for home aquaria, complete with reefs and clown fish. This demand began in the 1980s and reached a curious crescendo after the release of a animated film, *Finding Nemo*, in 2003. Shops and garden centres felt obliged to meet this growing demand by stocking their marine tanks with corals and crabs. Reefs were being raided for pets. Customers poured in, and then paid out with credit cards for a lavish array of tanks and pumps, plus a truly alarming battery of chemical solutions, just to keep their reefal microcosms alive.

These early hobbyists were a dedicated bunch. But they often paid a heavy price, not only in dollars but also in dead fish and rotting coral. It soon became apparent that the sustenance of a coral reef in the living room was not an easy matter. It required the maintenance of chemical stability. Water of good clarity; and water of high quality. The problem here was that aquarists wanted fish, lots of colourful fish. But fish meant food. And food meant faecal matter. And as you can no doubt guess— faeces meant trouble.

Faeces made matters worse because fish dung is rich in nitrogenous and phosphatic waste. Nor did it help that the hobbyist's local water supply was commonly rich in dissolved nitrate and phosphate from agricultural pollution. Little things like diatoms and Green algae living in the tanks loved all this waste because, for them, it was like living in a bucket of warm fertilizer. Hence, the tropical fish tanks of hobbyists would quickly develop a scum of Golden Brown diatoms, or a forest of Green seaweed. Something had turned their tanks into a Green Water ecosystem, and that something was faecal matter.

The aquarium trade therefore needed a remedy, to inhibit the 'algal problem' and return it to Blue Water heaven. At first, the remedy for green scum was the addition of various kinds of chemical solution. But the addition of chemicals became a disaster not only for the algae but also for symbionts trying to eke out a living within coral tissues. Little by little, the aquarium community began to realize that Darwin was actually correct. Ecosystems cannot be constructed from the top down, by starting with human whims and hungry fish. They must be built from the bottom up—by facilitating the growth of beneficial bacteria and algae.

These little benefactors would then, on their own, be able to help maintain the water balance and clarity.

Unfortunately for the hobbyists, all this meant adding a further battery of mechanical aids—fancy sumps for the nitrate-reducers; pumps for protein skimmers; and porous limestone boulders for algae and their invertebrate scavengers. Unhappily, it also meant waiting for many months before it was safe to introduce any fish. Or better still, doing without fish altogether, and making corals and sponges the objects of contemplation.

Maintenance of a balance within a coral reef microcosm can prove a tricky business, even today. Symbioholic corals, such as Elk Horn or Stag's Horn, are famously finicky—they demand low levels of nitrate and phosphate in the water, as well as stable and warm temperatures, and proper levels of illumination. They cannot endure profligate waste. Once those parameters for paradise have been achieved, it may be possible to enjoy a Blue Water ecosystem in the parlour. But there is often bad news just around the corner, too. Not a single part of this paradise can ever be allowed to slip. And nothing could be worse than to add too many fish, too many nutrients, or just a bucket load of silt and mud. Should any of that occur, then the system will quickly revert to a gooey mess, or a jungle of green and brown. The waters will return to a wasteland.

There is an even deeper truth here. This relates to the observation that any given ecosystem, such as a rainforest or a coral reef, will always veer towards the maximum rate of recycling of energy—be it sugars or phosphorus—within the system if it can. Indeed, some would go further and state that this maximal recycling is how an ecosystem can itself be defined.[7]

It is true that scientists such as Pamela Hallock of South Florida University had been warning about the dangers of high nutrient levels in modern reef ecosystems, ever since the early 1980s.[8] But fishermen were not listening because they were flush with money; the hotel industry was not listening because it was flushed with success. Hence the reefs paid the price. And those natural sinks for excess nutrients—unsightly places such as mangrove swamps and sea grass meadows—had already been bulldozed to make room for more hotels. Nobody was paying much attention back then. But today we are learning this lesson, as we read reports of great rafts of Green algae—mainly *Enteromorpha*—choking our coastlines as a result of nutrient excess.[9]

An answer to our question, therefore—how can a Blue Water ecosystem be sustained—has had to be learned by following Darwin's example: by talking with the hobbyists themselves—the pet shop philosophers.

Turning inside-out

As early as 1842, Darwin had also noted that his beloved coral polyps were highly sensitive to physical stresses, including excessive exposure to heat and sunlight.[10] But the first real signs that heat was becoming a serious menace came as late as 1979. By that time global temperatures around the world were seemingly starting to rise. Some corals were starting to look a bit pale about the tentacles too. And this 'bleaching' meant that their coloured endosymbionts had been expelled from the coral's tissues and into the surrounding waters. In other words, the symbionts had been obliged to sever their connections and live freely in the sea, if they could survive at all. That conferred on them, perhaps, some of the blessings of independence. But for the host corals, it was a symptom of sickness. Corals were to be the big losers here. They were starting to lose control of their harvest.

Within a few years, bleaching was starting to appear more frequently around the world. Corals such as *Acropora palmata* were turning white about the tentacles, then starving, then dying, only to leave behind great stands of dead coral, including those on the Pedro Bank. As you may have guessed, corals have not been alone in expelling their symbionts at these times, either. The giant protozoans we have met with, such as *Heterostegina*, have been caught bleaching as well. Not only that, but sponges and other invertebrates have also been turning pale. Many of these sickly organisms then get smothered by creeping mats of blue green cyanobacteria. Or they get replaced by waving meadows of nutrient-loving Green algae, such as *Halimeda*. One way or another, shallow water ecosystems in the Caribbean have been turning from Blue to Green.

During the 1980s, reefs were hit by yet another series of blows, not only white band disease but also by the death of algal grazers such as the sea urchin *Diadema antillarum*. And then came the added shocks of El Niño events and stronger hurricanes. One might think that there have always been hurricanes. But historical records suggest there have been whole centuries when hurricanes were rare.[11] That may explain why early colonists seemingly made no mention of them at all. Another rising threat has been that of human pollution and over-fishing. Under increasingly warm conditions, the natural bacterial populations have been able to reach epidemic proportions, infecting the coral epithelium and killing off the symbionts within. In this way, stands of living coral have been reduced from 50 per cent down to as little as 5 per cent over two decades. Major calamities befell the reefs in 1998, and yet again in 2005, during years of high summer surface temperatures. If the temperatures continue to get warmer, we are told, then the hurricanes will grow. And as the reef dwellers bleach, so the fish stocks will drop. If temperatures continue to climb at this rate, it is said, then all the coral reefs of the world will bleach and be dead by the end of the present century. The world of the reefs is being replaced instead by forests of fleshy algae. It is being turned inside-out.[12]

Here in the reefs, we therefore meet with an intriguing dilemma. On the one hand we humans can keep mining away at our mineral fertilizers in the mountains and flushing it out to sea. We can continue to treat our hotel complexes and the oceans beyond like a corporation sewage works. Such actions will turn the seawater green with plankton and weed. That might sound harmless enough. But plankton and weed will someday become so abundant that they will reduce the water clarity, making it harder for the coral symbionts to feed on sunbeams. The sea may, by then, have turned the colour of pea soup. Or worse, the colour of tomato soup. And the coastline will erode away, or drown beneath dung. We may call this 'the Green Water option'. It is, as you may perceive, a classic example of the kind of top-down thinking in which humans are seen as sitting at the top of the chain of command, and never mind what happens below.

On the other hand we could studiously ignore the fertilizer mountains and insist on recycling everything. Nothing need be wasted and little would then get dumped out to sea. In that way, we could treat our hotel complexes, and the oceans beyond, more like the coral reefs tend to treat themselves—by recycling, and regenerating everything on the spot. Such actions would indeed help to keep the waters blue, by lowering the levels of plankton. And corals would still be able to continue harvesting sunlight to feed their symbionts. The sea would stay blue. And the corals would thrive. We may call this 'the Blue Water option'. It is a kind of bottom-up thinking, recognizing the deep implications for ourselves of keeping things tidy at the base.

There are some signs that the Blue Water option could still be a winner. Each time the temperature has risen, it has also fallen back. Survivors have cultivated symbionts anew, usually more heat-tolerant strains. Clever forms of reef substrate have also been placed on the reef floor, to accelerate their recovery. New laws have been passed, as well, to protect coastal mangrove and reduce pollution by tourists. And programmes have been developed to educate the public. For this reminds us that the biggest enemy is not global warming but human ignorance. If ever signs are spotted that coral reefs around the world are turning permanently from Blue Water to Green Water ecosystems—and that those wrinkly films of wiggling cyanobacteria are winning the battle—then some will say that the endgame for reefs is near.

All this has happened before, of course. Many times over, as we shall shortly discover. The rock record is able to tell us exactly how this game can be won or lost, if we dare to look at the evidence beneath our feet. This same record also shows us which of the various ecological strategies are most vulnerable during the endgame, and how we humans may yet hope to survive. But for this we need a 'yardstick'.

Only connect

As we have seen, Robert Hooke long ago marvelled at the greening of his water butt, and the microscopist Robert Brown later explored the interior of his beloved green plant cells. And then along came Charles Darwin, enthralled by the blue waters of his coral reefs. So what are we to make, then, of this phenomenon—the Green and the Blue?

Each of our stories this far has been entangled with one of the biggest puzzles within Darwinian evolution—the secret history of the living cell. Each also presents us with a puzzle about the energy flow within a system. One of the great axioms of biology states that *all the energy flow seen in life takes place within and between cells*—in other words, within Hooke's secret chamber. But it is no easy matter to measure the energy flow within a big system, let alone inside a tiny cell. And the way in which 'energy' tends to flow through a given system is likely to vary very widely indeed. On the one hand, for example, this energy flow might appear at the outset to be rather smooth and predictable, much as is seen within Darwin's living coral reefs. On the other hand, the energy flow may well appear to be highly disrupted or even destroyed, such as during Lyell's troublesome mass extinctions.

So if we are to get to grips with the origins of the nucleated cell, we will need to take a very wide perspective on planetary ecosystems in time and space. We will need to take what is today called a 'systems approach'. That is to say, we should step aside from the top-down thinking of a Sunday School or even the bottom-up thinking of a chemistry class. For 'systems thinking' is almost like a third way of seeing. It sees the universe as nested and contingent, like a set of Russian dolls. All living things can be regarded as inter-connected. These connections are regarded as crucial to the way that living things behave.

So what do seawater colour, nutrient levels, and nested worlds have in common? More precisely, why do coral reefs prefer Blue Waters with low levels of iron, phosphate, and nitrate, whereas algal meadows like Green Waters a bit richer in dung? The answer, as it turns out, has to do with the flow of energy within a system, and the efficiency of that flow. There are parallels, here, with the flow of money both into and out of a human economy. Forget the modern world with its imaginary money for the moment, and think back to the Jazz Age of the 1920s. That was a time when the gold supply and the silver supply still really mattered. It was the flow of gold into the bank that helped to underpin the level of borrowing and lending that the bank undertook. If we stretch our analogy a little, we can say that the phosphorus within a living ecosystem behaves rather like the gold supply in an economy. And nitrate acts like its silver supply. Both control the level of ecological activity that can take place within that system. Too little of either, and the system starves to death. A steady supply of each keeps the ecology ticking over nicely. But if the system is flooded with phosphate or nitrate, then the ecology will be driven into cycles of boom and bust, as with the economy from 1920 to 1929.[13] Thus, if the phosphorus supply is restrained, there will be a rising need for greater efficiency and recycling, as in the 1930s.

There is more to it than this, of course. But it would be fair to say there is an average need for phosphorus atoms—in your body too—at the ratio of roughly 1 atom of phosphorus for every 104 atoms of carbon. Likewise for nitrogen at the ratio of roughly 16 atoms of nitrogen for every 104 atoms of carbon. This ratio of 1:16:104 is not entirely rigid in the real world, of course, but it typically approaches these values, as the American oceanographer Alfred Redfield first pointed out in 1934. The Redfield ratio is the Gold Standard of the living world.[14]

We can therefore say that phosphorus is to the energy flow of an ecosystem as bars of gold can be to the money supply of an economy. But how could we hope to measure the energy flow in an ecosystem, either living or fossil? Darwin must have pondered this puzzle while writing the *Origin of Species*.[15] His thesis, remember, was this. First, he pointed out that there is natural variation within any living population of a species—many types of organism have versions that are either big or small, pale or dark. Second, he calculated that creatures of a given species produce more progeny than can ever hope to survive—otherwise we would be wading knee deep through rats and rabbits on our way to work. After that, he conjectured that the lucky few survivors will owe their success to a modicum of greater fitness within their environment. Hence those that are less fit may not reach the age of reproduction—they are weeded out by the market forces of natural selection. And finally, he argued that a study of rocks shows us that the environment is not fixed; it changes continuously. And that means that the features defining fitness in any given creature must change alongside, just to keep in step.[16]

Taken together, these three processes—all of them involving energy flow—can be used to explain why organisms always seem to change from one place to another, or from one period of time to another. 'Transmutation by means of natural selection' is what Darwin called it. And Neo-Darwinian evolution is what we call it today, with the added perspective of genetics.

Can we ever hope to trace any of these patterns in the fossil record? Surely, those vast numbers of progeny have seldom been preserved for all time? That is mostly true. The embryos and juveniles have often lacked skeletons so that only a few have become caught up in the honey trap of the fossil record. There is worse to consider as well. How could we ever hope to measure the fitness of an extinct organism? That would surely mean putting numbers to things which are now lost forever from the fossil record, would it not? Well yes and no. Not everything has been lost. There are still a few groups in which meaningful variations might be measured rather well in the fossil record—fossilized shells like the giant protozoans we have already met. These Foraminifera, remember, are like the fruit flies of the fossil record.

Lines of communication

As we have hinted at so far, much of the energy flow in life is known to take place within cells themselves. Take a foram like *Archaias* from the Pedro Bank, for example. Photons from sunlight are used by its Green algal symbionts as a source of energy for the production of simple sugars such as starch. Conversion of that starch into glucose creates yet another energy-rich material. But glucose is no ordinary molecule. It has an almost Lego-like potential for the building of further complex organic molecules. One of these Lego constructions involves the substitution of phosphate. Within the mitochondria of the foram host, this would lead towards formation of ATP, a powerful form of energy storage.[17] And as soon as energy is released from this ATP by breathing in oxygen, it can then be used to help build materials such as enzymes and other proteins, from those recipes encoded along RNA, themselves controlled by the nucleus of the host cell.

This much about energy and internal economy was already understood. But a curious idea came to me during our cruise aboard *HMS Fawn*. I wondered whether the lines of communication within fossilized foram shells might help to provide us with a measure, not only of the size of its single cell, but of the flow of materials—and hence of energy—both within and around that cell. Such materials are likely to have included both the nucleus and the mitochondrion. In a very small cell, of course, a single example of each organelle might well be plenty. But in a large and complex cell, like *Archaias*, I mused, we might expect a more pressing need for multiple nuclei or multiple mitochondria, just to keep things alive and kicking across such a large space, especially when patterns of diffusion could be complex. Not only that, but in forams harbouring numerous algal symbionts, as does living *Archaias*, its guests would surely need to be shepherded into the light, but moved out of the way of other important activities, rather like children attending a school play. That being so, I wondered whether we could envision those curious pathways around the foram shell as a measure of the internal economy of the foram cell itself? In other words, could we use the architecture of different foram shells to measure the efficiency of 'energy flow', not only within living cells, but even within long extinct examples? If that were at least partly possible, then we might be able to hold a key that could help us to decode the history of 'energy flow' through life, across much of Deep Time.[18]

Now, the fossil record of Foraminifera is arguably without compare. Their little shells—more properly called 'tests'—span the fossil record from the start of the Cambrian explosion right through to the present day.[19] Such foram shells also occur in huge numbers, in all manner of marine sediments, from estuaries to the deepest layers of the ocean.[20] And their skeletons are complex—literally packed with information about the potential for energy flow within their living cells. Their shells might be able to provide us with a measure of metabolic activity through Deep Time. I was therefore eager to test if this was possible.

Before the advent of the personal computer, my initial attempts at modelling the potential of flow within the foram cell came in the form of weighed balls of plasticine—a rather sticky solution to a rather sticky problem. Later, I began modelling with paper and nautical plotting instruments. Both were used to test out the potential patterns of cell flow—more properly called cytoplasmic streaming—through the course of growth.

In this way, I tried coaxing out the rules for growth in giant protozoans such as *Archaias*. But, eventually, I decided not to attempt this by looking from the outside of the shell inwards, as most might expect. I needed to do it by looking at the problem from the inside out.[21] As soon as things were looked at in that novel way, it became possible to appreciate a curious fact. Many of the reef dwelling species we have just met were seemingly obsessed with one simple but very important problem in life—the problem of *communication*. The bigger the cell, it seemed, the more pressing was the problem of efficient communication within and across it. Why was that?

Of temples and tombs

Good communications, be they road routes, shipping lanes, or internet highways, are needed to ensure the efficient running of any complex system. The Roman Colosseum was a construction that allowed the maximum number of people to flow both in and out of a space without congestion. Not only that, but this flow needed maximum possible speed and the minimum possible lines of communication. The Colosseum provided a great spectacle for the public. And its design became a spectacular success. So efficient was this greatest of amphitheatres that it could, in theory, disgorge its 50,000 rowdy occupants in under five minutes. This huge achievement was made possible by its curiously short lines of communication. These lines were arranged in the form of radial corridors, circular corridors, and numerous stairways. Roman amphitheatres and their modern equivalents—our much-loved football stadiums, boxing rings, and sports arenas—are not just any old kind of space. Curious to report, they are examples of a deeply rational space.

At the opposite end of the spectrum in Rome can still be found many buildings whose aim was the exact opposite, to cut people off from the outside world—such as churches and cathedrals. Good examples of the architecture of isolation can be seen, for example, in the nave of St Peter's. Further examples can be seen in mortuary buildings, such as the Catacombs of ancient Rome. In each of these cases, we envisage a single rather narrow entrance, followed by a long walk towards the end of the chamber. Our procession from front to back can either be in straight line, as in St Peter's, or it can be circuitous, as with the Catacombs. But it will never take the shortest possible route, since the aim is not the efficiency of communication but isolation and awe.

From this brief inspection, it can be argued that we tend to encounter shorter and more efficient lines of communication in those places where commercial exchange and crowd control is needed, and longer lines of communication are found where protection and remoteness are the aim, places like temples and tombs. There are many buildings, too—ranging from mud huts to town houses—where the conditions may be said to lie somewhere in-between these extremes.

Suffice it to say that I was starting to think along these lines when writing a student textbook on microfossils back in 1976.[22] And I was starting to find this spectrum, ranging from short to long lines of communication, when examining the evolution of the foram cell and its shell. Time then to take a closer look at some of these beautiful patterns.

Communion

It happens that a good many objects can look much the same down a microscope as they do in daily life. A snail, for example, gets just a bit bigger; a finger nail looks a little more wrinkly; a grain of salt looks a bit more cubic. But a Foraminiferan shell can improve its appeal a good deal more than that. When placed in the palm of a hand, this shell can seem like little more than a chalky blob, if indeed it can be seen at all. But under a microscope, some of these shells can take on the appearance of a simple mud hut, while others can resemble a utopian city. Contemplation of this wonderful range in architectural form has provided delight ever since its initial discovery by Robert Hooke in 1665. Indeed, the coils of forams have even been seen as the inspiration for Art Nouveau in the 1890s.[23]

The key to growth in this group of protozoans is largely caught up within the living space of its shell. Like a human building, the material used for this shell varies from the gluing together of mineral particles—like a primitive mud hut—to the sophisticated secretion of chalky material around a 'former'—more like the casting of a concrete house. But what are the rules of the game here? With pencil and paper to hand, I therefore sat down and attempted to devise a few fundamental rules. These rules, which I called the MinLOC, or Minimum Lines of Communication rules, were briefly as follows: make chambers (rooms) that are wider than they are long; coil them around in a spiral so there is a short inner trackway (corridor or staircase); multiply the number of apertures (doors); make the apertures (doors) convenient for cutting corners.[24]

Many of the shells collected from the coral reefs—such as *Archaias*, *Borelis*, and *Heterostegina*—were found to compare remarkably well with the markets and circuses of Ancient Rome. Each of the little chambers in these foram shells is subdivided into even smaller rooms called chamberlets. Returning to our Colosseum analogy, we can see that it displays great passageways circling both around and within the building, to allow the various compartments to be connected efficiently. And much the same can be found within the fossilized foram shells, whose galleries are called canals. Returning again to the Colosseum, we can see that there was not just one but many dozens of doors that open out onto the street, to allow the crowds to exit rapidly, without fear of crushing and to discourage them from engaging in riotous behaviour. And much the same arrangement can be found within our colossal protozoan shells, where they are called apertures or more correctly *foramina*, meaning 'lots of little holes' in Latin. It is from these foramina that the Foraminifera—the hole bearers—get their name.

Recall that our travels around the Caribbean revealed that these 'Colosseum shells' were invariably owned by protozoans living in well-lit reefal conditions. So why was a complexity like that of the old Roman Colosseum needed within a single-celled protozoan shell? The answer to this conundrum, as we have already hinted, is that some of these protozoans are not really single-celled at all but rely on their cultivated algae to survive. Thus *Archaias* cultivates Green algae while one group of near relative likes to harvest Red algae, and another favours dinoflagellates. Both *Borelis* and *Heterostegina* farm Golden Brown diatoms—thousands of lodgers teaming within each shell.[25] Intriguingly, therefore, it was in *cells that had taken in strange lodgers* that I was finding the shortest and most efficient lines of communication. Put another way, it can be said that the Blue Water ecosystems in which these creatures live have seemingly encouraged a kind of cattle market for their symbionts, and hence a market-like construction for their shells.

Not all protozoans have shells of this complex kind. Some forams have constructed their shells along the lines of temples and tombs—with very much longer lines of communication. As with many temples and tombs, too, there is usually only one door to get in and out, no matter how big the queue, and even the rooms themselves consist of long narrow chambers, rather like the naves of Roman churches and catacombs. A fine example

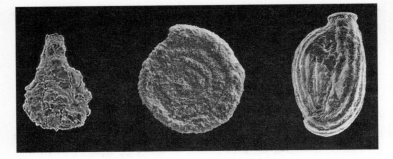

Figure 15. Three simple forams. *Saccammina* on the left has a flask-shaped shell made from mineral particles glued together. *Ammodiscus* at the centre has a coiled tubular shell constructed in the same manner. *Quinqueloculina* on the right has a shell like a cluster of bananas, made from chalky calcite. These three capture the first three steps of architectural evolution in this important group of protozoans. The field of view is about 1mm across.

of this principle is the shell of a creature called *Quinqueloculina*. Looked at from the outside, its chambers somewhat resemble a bunch of bananas (Fig. 15). This hexasyllabic bunch of chambers likes to thrive in physically stressed and very salty Green Water lagoons.[26] But from the inside, it consists of up to a dozen elongated cubicles arranged in an elongate spiral, rather like a maze.

Other kinds are found to thrive far below the photic zone, down on the deep sea abyssal plains. One of these is called *Saccammina* and has a shell that looks like an old brandy bottle (Fig. 15). It is roughly assembled together from grains of silt and clay, bits of mineral particles or sponge spicules. Yet another has a long tubular chamber, but coiled around like a Danish pastry, or indeed the horns of the god Ammon (Fig. 15), and is appropriately called *Ammodiscus*. Such tubular forms tend to flourish today in what we might call the environmental extremes—habitats that seem to be either very shallow (such as salt marshes) or very deep (such as the abyssal plains down to 6000m).

Intriguingly, when these three forms are lined up in a row from left to right, as shown in Figure 15, they exactly repeat the ancestry of those symbioholic forams we met in the reefs of Pedro Bank, such as *Archaias* and *Borelis*. For example, the fossil record shows that forms like *Saccammina* evolved in the early Cambrian, about 540 million years ago, while *Ammodiscus* evolved a little later, in the middle Cambrian, some 510 million years ago. Banana-shaped forms like *Quinqueloculina* didn't develop until the Permian, about 300 million years before the present. But last of all came *Archaias* and *Borelis*, appearing during the Eocene period, a mere 50 million years ago. We can be sure that this genealogy is correct for two good reasons: recent molecular studies confirm this pattern of development;[27] and microscopic examination of foram shells reveals the presence of ancestral architectures that have been repeated during the early stages of growth.[28] Indeed no living or fossil shell shows this story—from tiny blob, through coiled tube, towards complex colosseum—more clearly through the course of its growth than does one of my favourite little creatures, *Discospirina* (Plate 7).[29]

Discospirina is a rather glamorous protozoan. About the size and shape of a tiny coin, its shell builds outwards from a tiny blob at the centre. This is the first chamber—the proloculus in latin. It is found in every kind of foram, and its shape recalls that of their most ancient shared ancestor, *Saccammina*. Next comes a long spiral tube, like *Ammodiscus* but built from chalky calcite. This variant recalls that *Cornuspira* evolved from *Ammodiscus* in the Carboniferous, some 350 million years ago. Then appears a series of banana-shaped chambers—the *Quinqueloculina*-stage, first developed by ancestors in the Permian.[30] After this, the chambers flare out like a series of trumpets, in a manner that recalls the shell of its ancestor *Renulina*, which lived in the Eocene. Then comes a flourish of final chambers. Great arcs are followed by circles of dainty chambers, each one fussily divided into chamberlets that, when lit from below, give the impression of a brilliant starburst. These chamberlets mark the final adult stages of *Discospirina*, which first appeared in the fossil record during the Miocene, some 10 million years ago. While it somewhat resembles living *Archaias* in its architecture, the chamberlets are rather differently constructed. If we now stretch out all of these chambers into a

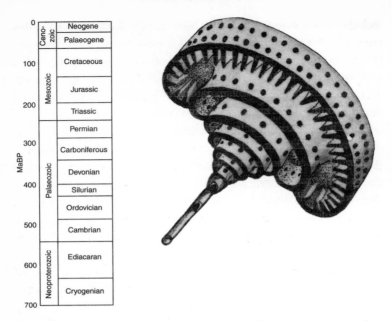

MaBP			
0	Ceno-zoic	Neogene	
		Palaeogene	
100	Mesozoic	Cretaceous	
		Jurassic	
200		Triassic	
300	Palaeozoic	Permian	
		Carboniferous	
		Devonian	
400		Silurian	
		Ordovician	
500		Cambrian	
	Neoproterozoic	Ediacaran	
600		Cryogenian	
700			

Figure 16. *Discospirina* has a shell that, during growth, recaps every stage of its evolution over the last 540 million years or so. In this image, these stages have been uncoiled so that they lie alongside the geological time frame. For a full image of the shell see Plate 7.

straight line (Figure 16), we can now see how *Discospirina* recaptures almost every stage of evolution in Foraminifera since the Cambrian explosion.[31]

Returning to less exotic examples, we can see how intermediate styles of architecture, lying between the extremes of *Saccammina* and *Archaias*, are commonly encountered in estuaries and sea grass meadows today. Some can even be used to reconstruct the history of sea grass communities through the last 50 million years or so.[32] Examples of this intermediate kind include protozoans whose shells can be said to resemble modern detached bungalows, each provided with a modest but adequate arrangement of a dozen or so chambers. Good examples of this type include the one shaped like a ram's horn that we met at the start of this book, called *Ammonia* (see p. 11). Another example includes the arrow-shaped shells of *Textularia*, simply made from mineral grains glued together. Both are known to flourish in the Green Waters of the Bocas del Dragos. But here comes the rub. Only *Archaias* and *Borelis*, at the end of this long lineage, are found to cultivate algal symbionts. None of the simpler or intermediate types are known to depend on them.[33]

It is now time to drive straight to the heart of the matter. We now find that it is the complex, Colosseum-like forams with their endosymbionts that are bleaching today from global warming. And it is the temple-like forams with their longer lines of communication that are sailing through this present crisis, as they always have in the past.[34] Complexity is all very well, but photosymbiosis can be a fragile and risky business. Complex constructions, after all, can have the furthest distance to fall.

How, then, have those symbioholics faired over the vast expanses of geological time? At long last, I was about to come face-to-face with this topsy-turvy world, to see if I could guess the name of the game being played. After a refit, our two survey ships *HMS Fox* and *HMS Fawn*, sailed northwards away from Trinidad—leaving behind the reefs of Tobago and sneaking past the volcanoes of St Vincent, to reach one of the most remote islands in the Caribbean: the island of Barbuda.

CHAPTER 6

WHEELS OF FORTUNE

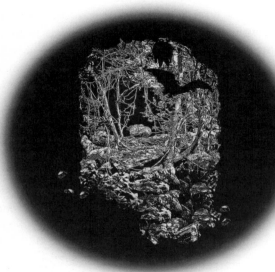

Cannibal coast

The dark mouth of an old sea cave beckoned. After sitting for hours astride an old grey nag in gruelling tropical sun, I anxiously let my bagful of rocks slip to the ground and crept towards the back of the cavern, half expecting to stumble on human bones. For this was a cave once inhabited by cannibals.

John White and myself had been trotting on horseback for miles across the island of Barbuda, at first along a ribbon of coastal sand dunes, and then plodding for yet more miles across a harsh landscape of limestone pavement.[1] Our heads were boiling from solar rays and our bottoms were sore from chafing saddles. But our horses were even more fretful because their feet were unshod. For us, the limestone pavement was an object of curiosity, but for them it was an ordeal, sculpted into a rockery as brash as broken glass. It felt cruel to urge them on. But urge we did.

Vegetable growth along this Atlantic coast of Barbuda—sword-leaved agave, turk's head cactus, and prickly pear—could seem as spiky and unforgiving as the rocks. But now and then, we would come across little tokens from a world long vanished. Our first token was a scattering of large snail shells that had somehow found their way up onto this limestone pavement from the boiling sea below. They consisted of Queen Conch shells—*Strombus gigas*. That was not surprising because the back reef here was part of their breeding ground. But there was something strange about their appearance. They were surprisingly large. Some were as big as my foot. And instead of being pearly and pink, like living *Strombus*, they were as chalky and fragile as old Amerindian bones. Curiously too, the spire of each shell had been punctured by a neat round hole, showing how they had been attacked with an axe made from the spire of another conch (Figure 17). That was a vital clue, as later shown to me by Amerindian archaeologist Dave Watters of the Carnegie Museum in Pittsburgh. During a subsequent month-long expedition to Barbuda in 1984 alongside Dave Watters and Jack Donahue, we came across a mound of such *Strombus* 2m high and an incredible 7km long. Carbon dating showed this shell beach had been there for some 3000 years.[2] These bone white shells, shell mounds, and *Strombus* beaches were the remains left behind from the feasts of long lost tribes of Amerindians— Arawak fishermen, and Caribs or Cannibals as they came to be called. It was a startling scene to stumble on, untouched by time.

The second sign of change, as we plodded along this coastal track, was more subtle. We kept tripping across isolated turban shells, each pleasingly marbled with black and white. These were the remains of *Livona pica*, a marine gastropod that lives today around the crests of the

Figure 17. Reconstruction of an extinct rice rat, of the kind found in the caves of Barbuda, here standing alongside a queen conch shell. The rounded holes in this shell were made by Amerindians, puncturing it with the spire of another shell to release the adductor muscle of this marine snail and get at the meat inside.

fringing reef. Strangely, these snail shells had found their way inland not from the activities of Arawaks but on the backs of generations of hermit crabs. It seemed that land-dwelling crustaceans had carried them inland from the beach, possibly over hundreds of years and sometimes for many miles.[3] But the oddest thing here was their size—almost twice that of examples along the shore today. Taken together, the landscape was painting a picture of visible environmental change. Many reef-dwelling creatures such as *Strombus gigas* and *Livona pica* had been getting smaller and scarcer with time.

We pondered these symptoms of reef decline as we picked our way slowly and painfully across the narrow limestone pavement, caught between old sea cliffs on the left and raging surf to our right (Figure 18). It was an unsettling place to work. While skirting along the foot of these ancient sea cliffs, it was easy to imagine being Arawak people hiding from raiding parties of Caribs; to imagine the boom of the reef as the sound of Carib tom-toms and the cries of seabirds as the screams of native children. Above us, too, rose the Barbuda Highlands—a remote plateau replete with swallow holes and jungle creepers. It was at this

Figure 18. The writer exploring with unshod horses along the remote eastern coast of Barbuda in 1970. In the foreground, the horses are grazing on the old reef platform, left behind by retreat of the sea 125 thousand years ago. In the background are the ancient sea cliffs and caves of the Highland Limestone, maybe a million years older.

point that we came across the gaping mouth of this old sea cave. And it was here that our most serious symptoms of change were to be found.

Caverns

As we crept towards the back of the cavern, past hanging roots and creepers, our eyes grew adjusted to the gloom, and the roar of the reefs became little more than a distant hiss. Back here in the cool, the floor was covered with a thick carpet of black and crunchy soil—not of bones but of dung. Millennia of bat droppings. Rooting about in the dirt here, we knew that a previous visitor had discovered the bones of an extinct Caribbean rodent which had lived in the caves some hundreds of years back. These giant rice rats—called *Megalomys*—were almost the size of cats. They had disappeared from the region some time after the arrival of European colonists, perhaps as late as the early twentieth century.[4] It was while contemplating this loss that we spotted, for the first time, some clear evidence for reef extinctions, plastered along the walls of the cave itself.

Over previous months we had been sampling and studying the living reefs and lagoons, mangroves, and algal mats around this unspoilt tropical island, and so we knew the biota fairly well. Along the coastline we had found great cathedrals of living Elk Horn coral, and shoreline sands speckled with strawberry pink shells of the Foraminiferan *Homotrema rubrum*. Taking a step or two inland, however, we could climb up onto an ancient wave-cut platform and find the remains of exactly the same organisms embedded within raised beaches some 125,000 years older, now entombed in solid limestone.[5]

Inquisitive as ever, we climbed a few steps higher in the cliff and found further signs of eroded reefs and yet more signs of fossilized beach sand. The higher we climbed, it seemed, the older and more bleached these deposits became. Some of these higher reefs are thought to be about 400,000 years old.[6]

These rising steps of reefs were the equivalent of that staircase of sunken shorelines we had found around the edges of Pedro Bank in the months before. Those sunken shorelines, remember, provided evidence that sea levels were once much *lower* than today. Along the cliffs of Barbuda, however, we were finding evidence for several episodes of sea level that seemed much *higher* than today. With all these ups, and a nearly equal number of downs, we could see that coral reef platforms around the Caribbean have been colonized and evacuated some dozens of times, as sea levels have either risen or fallen in response to the waxing and waning of Antarctic ice caps.[7]

Even with all these comings and goings of reefs and sea levels, however, we could at first find no signs of any conspicuous extinctions in the shelly biota. That was to change, though, as we peered closely at the walls of these old sea caves. Looking at the deeper layers here was rather like peering inside a birthday cake, instead of staring blankly at the icing on its top. Here, in these inner layers, we could at last see fossil shells unlike any found living around the island today.

These strange fossils somewhat resembled, in size and shape, a collection of rice rat's eyes. Clusters of eye-shaped fossils were staring out at us from the pale grey limestone (Figure 19).[8] Closer examination, and much further research, showed them to be calcareous shells. They were the remains of an extinct Foraminiferan that Joseph Cushman had been the first to describe, which he called *Amphistegina rotundata*. From its shape, size and ecology, it was arguably a symbiont-farming protozoan—yet another symbioholic—that had managed to reach rock-building proportions around the Caribbean Sea. But it has only been found in rocks about 3 million years old, laid down when the region was even warmer. No living examples are known.

A further clue to this turnover in reef species some 2.7 million years ago was provided by yet another fossil that we found within this rock—a kind of fossilized stony alga allied to *Aethesolithon*. With cells like fancy lace, this calcareous seaweed had seemingly originated in the Pacific ocean.[9] This, and much other evidence besides, suggests that Caribbean reef communities suffered extinction when North and South America were conjoined by the Panama isthmus some 3 million years ago. In this way, the Caribbean sea had become isolated from the Pacific ocean to the west. No longer, then, did the warm waters of the Caribbean flow unimpeded westwards into the Pacific. Instead, the newly uplifted land around Panama blocked the free passage of the westerly flowing currents. The warm waters of the Caribbean were thereby made to curl back on themselves, past the panhandle of Florida and up towards the chilly North Atlantic. In this way, the balmy Gulf Stream was born. And that meant that its balmy waters were sailing into cold Arctic regions, and falling on neighbouring lands as rain, and then as snow, then forming ice, and then ice caps. In such a way, the great cycle of ice ages in which we now live is thought to have begun.[10]

Figure 19. Wheels of Fortune. The walls of the old sea cave contained the fossils of a now extinct reef ecosystem, over a million years old. The thick coiled shell of foram *Amphistegina rotundata* (above) may have acted like a fibre optic lens, to bring light to symbiotic diatom algae living inside its single cell. This protozoan once clinged to an extinct Red alga here called *Aethesolithon* (below) that shows its lacy network of polygonal cells. The field of view is about 10mm across.

A lost monk

Our hike along the cannibal coast had therefore brought us face-to-face with several lines of evidence that pointed towards major changes in climate, in ecology, and the overall biota. Not only had reef-dwelling creatures suffered decline and extinction in the distant past, but more recent communities had been facing the roller-coaster of sea level changes too. Latterly, cave-dwelling rodents and Arawak Indians had even been driven to extinction. Since the arrival of scuba divers and intensive fishing, all manner of fish and shellfish have been declining in size.[11] It is here that we come to a worrying part of our tale—the risk of collapse faced by coral reef ecosystems and their symbionts today. And a good place to begin this story is not with corals, but with the mammals of the reefs.

At one time there were large colonies of the Caribbean monk seal—*Monachus tropicalis*. This large but shy sea-going seal was known to have dined upon a diet of goatfish and grunt, captured around the coral reefs of the Caribbean. Monk seals also used the sandy beaches around those isolated coral islands called cays for hauling-out and breeding. And breeding, it seems, was something monk seals were once doing in huge numbers, until something went badly wrong.

For centuries, we are told, these monastic beasts were being clubbed to death by fishermen, and their blubber boiled down for cooking oil and lamp fuel. Nothing much went to waste. Their skins were either turned into oil skins or into handy carrier bags. No one thought any of this much worth worrying about because there were, reputedly, millions of monk seals.[12] And indeed there were—in the eighteenth century. But by the nineteenth century, the monk seals could no longer rise to the frenzied demands being made upon them for skins and blubber. Rather unexpectedly, their numbers began to dwindle, steadily at first, and then very sharply indeed. The last seal was seen sitting like a mermaid on Seranilla Bank, just to the west of Pedro Bank, in 1952. Not a single monk seal has been seen since.[13]

From this sad story, it seems likely that the Caribbean monk seals basked in their tens of thousands around the shores of Pedro and Barbuda during the last ice age. Their doom cannot have been helped by a reduction in habitat occasioned by the melting of glaciers and the rising of sea levels. Both changes inevitably caused fewer breeding sites. But post-glacial sea level rise cannot have been the main cause of their demise. We can be reasonably sure of this because they had survived both climate and sea level changes for some 15 million years beforehand. Dozens of ups and a nearly equal number of downs in sea level. Most likely, therefore, their demise was precipitated by something new. And the money is on the impact of a new predator on the ecosystem—the European colonists.

This story is not singular. It has been repeated again and again. We could instead choose the cases of the manatee, the queen conch or the green turtle. All three are highly specialized and efficient feeders, mainly upon sea grass meadows. All three were so prolific at the time of Columbus that they astonished early explorers. And all three are now locally extinct and in global peril, following the adoption of the face mask, flippers, and scuba diving during the 1960s.[14] In other words, the food web of reefs has been collapsing since the mid-nineteenth century. If so, then we might not be at the beginning of a mass extinction, as often maintained. We may be hurtling towards its conclusion.

The casino of doom

Some big creatures like the turban shells we met with can get smaller with time. Some like the monk seal can even disappear. But is it possible that big and complex ecosystems can really collapse completely? Indeed, could it even be argued that complex ecosystems are more vulnerable to collapse than their simpler relatives?

To explore this dangerous wheel of fortune, we need to enter something like a casino. Unlike the Bahamas, though, Barbudans do not own up to a casino.[15] But they are not averse to a good game of cards. So let us sit out on the veranda and deal ourselves a hand of cards. Reading the story of life from the rocks is, after all, a bit like sitting down to a game of cards. Except that *we do not know the rules of the game being played.*

Consider, for instance, the four hands of cards shown in Figure 20. It might seem a fair bet that the royal flush, at top left, is the winner. Or that the run in spades, shown at bottom left, is a good bet. But what if the game being played here is not poker or cribbage? What if it is a game of Fibonnacci?

Now the Fibonnacci series—which runs 1, 1, 2, 3, 5, 8, 13, 21, 34, and so on, with each number being the sum of the previous two—is a code enshrined within much of the living world. It contains the rules for growth and packing of cells, from which good order can best arise. It is

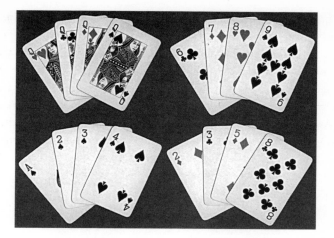

Figure 20. Reading the history of life in deep time is a bit like playing this game of cards. The challenge is to guess which is the winning hand here?

essential to life. If so, then it is the hand of cards in the bottom right of Figure 20 that wins the game here.

As we sit down to the table, we therefore need to ask 'have I got the right number of cards?' and then 'what cards have I got?' But most of all, we really need to ask 'what game are we playing?' That might seem sublimely obvious to us now, but with the fossil record, we are never told the name of the game that life itself has been playing. The universe remains resolutely mute upon this matter. So we must sift through the history of life in the rocks to seek for patterns in fossils through time. From thence we will need ingenuity, and a portion of lateral thinking, if we are to work out the name of the game that life has been playing.[16]

The Danish philosopher Søren Kierkegaard is reputed to have said: 'we are condemned to live life forwards, but can only understand it backwards'.[17] And if that saying turns out to hold true for our own lives, then it will turn out to be especially true for the grand history of life itself. To understand this, we now need to take a closer look at the great tree of life. And for that we will need to get across the lagoon.

CHAPTER 7

MANGROVE TREE

Frigate Bird

After splashing across the salty expanse of Codrington lagoon, John White nosed our dory craft towards a thicket of mangrove. We then switched off the outboard motor, to let our little boat, and our heartbeats, slow down. Soon enough, we were nosing gently through the dark green glade and punting along with a paddle. It was strange to feel the tickle of mangrove leaves on our arms after hours spent in open water. Deep inside the swamp, the water looked and smelled rather curious too. Elsewhere, the brine had either been a limpid blue colour, as in the reefs nearby, or a soupy jade, as in the lagoon we had just traversed. But here in the mangrove swamps of Barbuda, the water more resembled a brew like dark Guyana rum.[1] One could almost smell the alcohol, its sweetness bubbling up towards us from the peat below. This vapour wasn't exactly

Nature's hooch, of course. It was bacterial burps, including Nature's methane.

After several moments peering into this fetid swamp, we glanced upward to find ourselves almost eye-to-eye with a family of 'pterodactyls'. Or at least, that is what they at first resembled. With beady black eyes, sharp beaks, and iridescent plumage, they were flapping about near the top of a mangrove tree, sitting in a construction that looked like a badly built bonfire—with sticks poking out at jaunty angles. We had sailed into a nest of big baby frigate birds—*Fregata magnificens*.[2]

These tropical frigate birds have a wingspan greater than any other avian of similar weight—over 2m in length—and they often choose to stay aloft for weeks at a time. For all that, they are somewhat misnamed, for they could clearly have outmanoeuvred any naval frigate during combat at sea. So agile are they, that they routinely snatch flying fish while on the wing. Frigate birds are famous for piracy too, swooping down like kamikaze pilots, forcing booby birds to drop their lunch and flee for safety. The attackers then dive down and gobble up the stolen meal, hopefully before it dissolves into clam chowder in the sea.

Barbuda lagoon therefore felt like a perfect hideaway for such prehistoric rascals. Hidden away from pounding waves, convenient for quick bombing raids, and many miles from the nearest human habitation, these birds could feed and breed in peace. It is not the frigate birds, though, that are important for our story. It is the home in which they chose to build their nest. For it is the *mangrove tree* itself that arguably holds a clue to the hidden history of the cell. This clue has nothing to do with being stuck up a creek with just a paddle; and it had everything to do with the architecture of the tree itself. Let me try to explain.

The cactus Tree of Life

The concept of a Tree of Life is very old indeed. First made famous by Charles Darwin in the *Origin of Species* in 1859, it has antecedents that stretch far back, through the oak groves of the ancient Druids, via the musings of Aristotle, to the Tree of Life in old Babylon. In quite a few cases, we know what these old trees looked like because they have been carved into stone, depicted in paint, or even woven into textiles. Interestingly, all of these very old images, and those of Darwin's sketches too, take on a rather familiar shape.[3] It is the kind of tree a child might choose to draw. There are no roots to speak of. A short tree trunk therefore rises directly out of the ground, and the branches ramify fruitfully into the sky above.

For a variety of reasons, including the cultural ones mentioned above, this charming old Tree of Life has blossomed into an icon for tracing both plant and animal ancestries in deep time. Hoorah for all that, of course—for the tree is one of evolution's greatest assets, serving as its dearest logo. But as we shall explore below, such a simple tree also presents us with an interesting problem. Or, indeed, with a set of problems. So thorny is this logo of evolution that we shall call it the Cactus Tree of Life.

The Cactus Family is just one of hundreds of flowering plant groups native to the central Americas. Cacti are not really rainforest plants, of course, although there are a few, such as *Epiphyllum*, that have made the jungle their home. They are best adapted to conditions of low rainfall, such as those found in deserts that stretch from Mexico down to Chile, or around low-lying coral islands such as Barbuda. Some, such as the turk's head cactus, take on a simple form and grow upwards into a single stem that is shaped a bit like a barrel. As with all cacti, the leaves and branches have been reduced, through aeons of evolution, until they are little more than tough and thorny spines. Many cacti tend to look like this. In the turk's head cactus, however, the generative zone forms a distinctive and amusing little topknot. Admittedly, the whole ensemble doesn't much recall the head of any Turk we might expect to meet today. It does, however, resemble the turban of an old Ottoman Turk. By expanding its stem, and reducing its leaves to spines, this cactus has found a way to cut down on water loss by transpiration—a harrowing problem in hot dry winds like those that blow across Barbuda in the summer. In addition, these cactus spines help to deter thirsty animals from stealing the water stored within specially adapted cells that lie deep inside the stem.

Much more common is the famous prickly pear tree, *Opuntia ficus-indica*. This cactus is so-named from its pear-shaped sugary fruits, which can turn as pink as a rose when they are ripe for eating. So spiky is this plant, however, that it is more usually employed to keep unwanted animals out of vegetable gardens. At the time of our visit in 1970, both goats and donkeys were still allowed to roam freely around Codrington village, so that a cactus hedge was used to help keep the asses out of gardens. It has to be admitted, alas, that a hedge of prickly pear was sometimes less than successful for this purpose. On one occasion I caught a donkey browsing in our back yard and, on a whim, decided to climb aboard him and hitch a ride. The donkey looked up somewhat bemused, and carried on munching. Then it started to trot forwards gently. Then forwards rather firmly. Then very firmly indeed. And then, to my horror, he charged at full tilt, only halting as soon as we reached the wall of *Opuntia ficus-indica* at the far end of the garden. The donkey then swung its rear end through the air, delivering me with great precision onto a wall of cactus needles—to the lamentable satisfaction of my colleagues. It was my first bruising encounter with the Tree of Life.

Opuntia might be thought to provide a reasonable analogy for the Tree of Life not least because the prickliness of its branches charmingly matches the prickliness of some of its advocates. But it provides a pleasing analogy for another reason, too—because of its distinctive shape. With almost no roots to speak of, *Opuntia* first produces a single 'branch' that rises up from the ground. The lowest branch on the prickly pear cactus can serve as a symbol for our bacterial ancestors at the base of the tree here, much as argued by Huxley and Haeckel. From this first branch arise two new branches, which can stand for single-celled protozoans like the Foraminifera we have met in the reefs and caves of the Caribbean. These, in turn, branch yet again to produce the algae and land plants on the one hand; and fungi plus invertebrates and vertebrate animals on the other. These top branches just referred to can be called 'crown eukaryotes' because they reside in the crown of the tree itself.

This is more-or-less how the tree of life looked when I was a student. This cactus tree looks rather easy for us to understand as well. Interestingly, though, it has turned out to be rather misleading as a model for evolution. The tree of life, it seems, may not have been shaped like a prickly pear cactus at all. Instead, it may have been shaped rather more like the Mangrove tree we met in the swamp. Time to sail back across the lagoon and take a closer look.

The Mangrove Tree of Life

Peering again at those Mangrove trees on the far edge of the lagoon, we can see that each comes provided with a distinct crown of branches, bearing large leathery leaves. It was in the crown of these trees that those frigate birds liked to build their nests. Beneath this crown stands the stem of the tree, which can be seen rising out of the water. Looking more carefully, however, we can spot that there are a number of other branches—called aerial roots. These descend from the outer branches, to go plunging through the water and into the sediment below. If we snorkel beneath the water line, we can also see that the lower part of the mangrove tree is not feeble like a cactus. It comprises a branching mass of roots, as ramified as the crown of the tree above. This pattern—massively branching roots below, and interlinking stems feeding up into the branching crown above—closely mirrors our second way of looking at the Tree of Life. Out of reverence to *Rhizophora mangle*, we will therefore call this the Mangrove Tree of Life (Figure 21).

With the mangrove model, the roots of the tree are very large and visible—standing for what we now understand about the various bacterial lineages. These are known to contain more genetic and metabolic diversity than that seen in the whole of the eukaryote crown above. Above the bacterial roots rise multiple stems—somewhat resembling the aerial roots of the mangrove—leading towards the eukaryote crown of the tree above, including those groups we have met with, such as the Foraminifera, dinoflagellates and algae, Green, Red, and Brown.

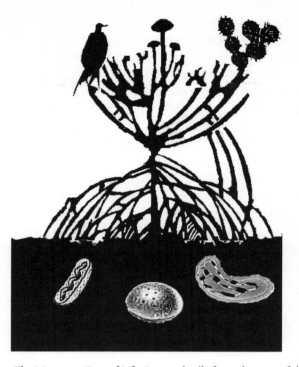

Figure 21. The Mangrove Tree of Life. Bacterial cells form the root of the tree. Animals and plants (including forams) branch off the crown near the top. But different groups of bacteria can be seen to contribute organelles to the crown from the roots below. The mitochondria seemingly arose by a host taking in oxygen-breathing proteobacterial cells (at bottom left) as symbionts. The chloroplasts of algae and plants arose later, by a host taking in cyanobacterial cells (at bottom right) as symbionts. The origin of the nucleus (at bottom centre) is more controversial.

Clearly, what we have here is a Tree of Life. But it is not shaped like a prickly pear cactus at all. It is shaped more like a mangrove tree. This very modern view has some very important implications. It means that the eukaryote lineages did not rise upwards as a single stem, from a single group of bacteria down below. Instead, it implies that the lines have got mixed up. Hybridization has been taking place. Eukaryotes are not a single entity. They are a kind of symbiotic chimaera, like the Sphinx—part one thing and part another—in which different kinds of prokaryote have taken to living together, permanently. In this view, as we shall see, one kind of bacterium may have given rise to the host cell. Another has given rise to the mitochondrion. And a third—in the algae and plants—has given rise to the chloroplast. Seen in this way, each eukaryote cell can be called a chimaera. But not merely a chimaera on the outside, like the Sphinx of Egypt. It is a chimaera on the inside. We here call this *the Riddle of the Sphinx Within*.

How did scientists ever come to such a strange conclusion—the Riddle of the Sphinx Within? As we shall see, the evidence can still be found hidden deep inside the cells themselves, including our own. And it has come forth in many forms. But the decoding of this curious chimaera has had—like the history of the little cells themselves—a very long and tortuous path.

The trouble with lichen

Perhaps the first hint that all might not be well with that old Cactus Tree of Life came from the study of lichens. Of all the curious life forms that Robert Hooke had marvelled at, when peering down his microscope around 1665, there were two primeval forms of life—Green algae and colourless fungi—that seemed especially difficult for him to understand. It just so happens that these are two of the groups that are typically found living together within lichens (Figure 22). As we have seen, Hooke was particularly puzzled by the ability of both groups to emerge 'spontaneously' upon the surfaces of old brick work and on stone walls. Seemingly, they were able to grow out of little more than a mixture of air and dirt. Very puzzling.

Lichens, like dirt, get everywhere. They thrive along the margins of arctic glaciers and range right down into subtropical deserts, any place

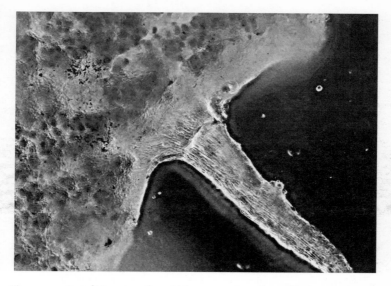

Figure 22. Just living together. Lichens consist of symbiotic associations between algae and fungi. In this 'dark field' image, raspberry-like clusters of single celled alga *Trebouxia* (dark spots at left) are hiding beneath a protective blanket of fungal filaments. The algal cells therefore appear as though seen through a wet shower curtain. Fungal filaments also combine to form bundles of root-like structures (at right) to help extract nutrients from the rocks. The field of view here is several millimetres across.

indeed where water is scarce and rocks can be found in a raw state. In more familiar settings, old graveyards and tombstones can often be found carpeted with their curious heraldry—crusty festoons of yellow, white, and black. In the worlds of Medieval to Renaissance scholarship, still much bemused by astrology and magic, it is no surprise that lichens provided writers—including William Shakespeare—with evocative symbols for love, death, destiny, and the passage of time. One of my favourite stanzas on this is comes from a Saxon poem describing decaying Roman baths in southwest England: 'Often this wall, grey with lichen, and red stained, unmoved by storms, has survived as kingdom follows kingdom follows kingdom'.[4] But it was not only poets who lauded these growths of lichens. Some of the earliest scientific illustrators paid careful attention to their crusty festoons as well. In this respect, it is interesting to recall a curious fact which came to light quite recently, when an old Royal closet was being cleared out at Windsor Castle. Quite how this cupboard had resisted a proper spring-cleaning for so long isn't explained. What has been made clear to us, however, is that this long-delayed dusting resulted in the discovery of no less than 27,000 drawings and paintings, all made in the 1620s.[5] And it was among these that some of the first ever drawings of lichens emerged. But it wasn't in Windsor that the lichen puzzle was solved. It was in Switzerland.

In 1867, Simon Schwendener proposed that lichens were not simple or single organisms at all.[6] They were associations between different kinds of living organisms—symbioses of algae and fungi that each gained a mutual benefit from their association. This provided some of the first indications that biological consortia can add up to something that is much more than the sum of its parts. This grand idea, that lichens were not simple creatures at all, but algal-fungal consortia was enthusiastically followed up by a keen naturalist better known for her children's stories—Beatrix Potter. Unhappily for Ms Potter, it was the right idea at the wrong time. Her work on lichens was rather coolly received.

Today we can readily observe such biological communes with our microscopes and molecular tests. Such studies confirm that lichens are neither simple nor single organisms at all. There is no individual here. Instead, most lichens are built from deals struck between various kinds of fungi and a Green alga called *Trebouxia*.[7] The fungi perform their usual trick of helping to dissolve rocks, minerals, and woody materials to release nutrients. Their algal guests profit directly from those hard-won nutrients and also gain protection from desiccation within the tough cell walls of the fungi. The algal cells of *Trebouxia* then release photosynthetic sugars back to the fungal host. It seems that this alga is not invariably dependent on its host and some strains can live out their lives alone. And unlike some of the coral and foram symbioses we have seen, the fungal host can also live without its symbionts. Together such consortia are, however, able to occupy niches that would otherwise be denied to each participant alone, such as rugged cracks within rocks on dry land. There is a conspicuous division of labour here. But the sum is much greater than its parts. In Darwinian evolution, that is a winning combination.[8]

A Soviet tale

Lichens and their curiously communal story did not really take root in the scientific literature for several decades. Communes in biology were perhaps regarded by the scientific establishments of the day as strictly off-limits. That is, until a scientist in pre-Revolutionary Russia took a rather courageous step. Constantine Mereschowsky was a Professor of Botany at Kazan University. Back in 1905, he had the audacity to argue that the chloroplasts found inside algae and plants were not just handy gadgets found within their cells. They were at one time free-living cyanobacteria—organisms in their own right.[9] Later, they became incorporated as symbionts, rather like the algae we have met living inside corals and Foraminifera—although rather little was known about those symbioses then. And finally, they became permanent guests—as photosynthentic organelles.[10] It is no surprise to learn, of course, that Mereschowsky's ideas fell largely on deaf ears. His tales of consortia sounded—as had Schwendener's lichen story before them—rather like the thoughts of a meddlesome Marxist philosopher, or like a dangerous dream.

All this was to change after the Russian Revolution in 1917. Biologists in Russia became more receptive to communal stories about cells. Thus it came about that a botanist called Vladimir Kozo-Polyansky felt emboldened to take the curious ideas of Mereschowsky a whole stage further. He pulled together a huge body of work to argue that chloroplasts were not the only organelles that had arisen from symbiosis. Mitochondria had been engulfed in this way. So had the nucleus. And so had the flagellum. All this was inferred from the minutiae of living plants and algae. It was a great leap forward in biology. He even called it a 'new principle of evolution'. There was, however, one niggling little problem with his great synthesis. He was too far ahead of his time.[11] Although he published a seminal volume on Symbiogenesis back in 1926, few within Russia would promulgate it, and even fewer outside of Russia would get to hear of it. Only in 2010 did the first English translation of his work become available.[12]

There are some interesting lessons to be learned here. If a hypothesis is correct, and *if scientists are allowed to explore the world freely*, then more than a single scientist will stumble upon across a major truth at some point in time, like ants bumbling across a forest floor. That is, for example, how Alfred Russel Wallace stumbled upon the self-evident truths of natural selection—twenty years or so after Darwin had first perceived the outlines of this line of thought. A constant process of looking and thinking, aided by the development new techniques—will usually bear fruit. And that is more or less what happened with the symbiotic theory of cell evolution outside of Russia. Key steps were being made on the development of the electron microscope by RCA within the United States during the 1940s. Then followed X-ray crystallography, and the structure of organic compounds, quickly followed by the decoding of DNA by James Watson and Francis Crick in 1953. And then came comparisons between RNA sequences in different kinds of organisms, like those made by Carl Woese of the University of Illinois, in 1977.[13] These and a myriad other advances, helped to roll back the curtain. But a proper appreciation of this tale requires that we take a visit to Amherst.

Tying the knot

'When visiting Amherst in Massachusetts, you *must* arrive in the Fall.' Or so I was told, and it is easy to see why. For this is the place where homes of American literati can be found hidden among leafy glades. And Fall is the time when leafy glades wear colour combinations no self-respecting person would ever dare to sport—even in the privacy of the boudoir—purples with yellows, and reds with greens. Amherst, it seems, is one of those places where chlorophyll finally goes to heaven, after a short life spent reclining in the sunshine.

But Amherst is not just a strikingly colourful town in the Fall. It is home to two rather famous women. Indeed, I once found myself sitting in the kitchen of one of them, and gazing out across at the garden of the other. That garden at one time belonged to a rather secretive poet called Emily Dickinson who lived and loved in vain here during the mid-nineteenth century. She wrote many poetic gems but a favourite one of mine goes something like this: 'Faith is a fine invention, when Gentlemen can see, but Microscopes are prudent, In an Emergency.' Emily Dickinson is famous, of course, as a philosopher of the soul.

More to the point though, the kitchen in which I was sitting belonged to a woman who helped to change the way we think about the deep history of the cell. She had for long been amassing evidence to show that the eukaryote cell is not really a single entity at all. That both Mereschowsky and Kozo-Polyansky were right. The eukaryote cell is a permanent symbiotic association between several different kinds of bacterial cell. Each little bacterial worker then brings something special to its larger commune—protection, organization, locomotion, sugar, and energy—to the point where this commune can no longer be dissolved. Symbiosis has turned into symbiogenesis—the genesis of a wholly new grade of organism through the merging together of symbionts.

Not surprisingly, this viewpoint sparked a great deal of controversy at first. It tended to challenge the ways in which we thought about our place on nature. And even about how nature actually works. For I was privileged to be sitting in the kitchen of Lynn Margulis. And Lynn was not merely a philosopher of the soul: she was a rarer species—a philosopher of the cell.[14]

Any visitor to the home of Lynn Margulis was likely to be struck by the ways in which it worked. Her old mansion contains a warren of chambers and corridors. While I was there, her home played host to a constant stream of visiting scientists, researchers, students, and even an old stray dog. She even cultivated us like algal symbionts, feeding us up and shepherding us around. The lively movement of occupants through her house—including myself—seemed curiously poetic, too. It called to mind the process of cytoplasmic streaming within a eukaryote cell. It is time, then, to look more closely at the Margulis phenomenon.

Queen bee

The Late Lynn Margulis has often been regarded as the Queen Bee of Microbiology but she began as a young woman in a hurry. She began attending lectures at Chicago University from the age of 14.[15] By the age of nineteen, and rather too soon perhaps, Lynn became the first wife of budding astrobiologist Carl Sagan. By 1959, she was tracking the genetics of cell organelles. In no time at all she was causing waves with the new evidence from microscopy. By 1966, she had the germ of a grand idea, but her early drafts were rejected by no less than fifteen journals before an acceptance was received from the *Journal of Theoretical Biology*.[16] And even then, she had a battle on her hands. Her thesis—now called the Serial Endosymbiotic Theory—if we boil it down to its essence, runs broadly as follows.[17]

Each of the organelles we have met with thus far comes provided with its own walls—membranes like those enclosing the cell itself. And several of these organelles, namely the nucleus, the mitochondrion, and the chloroplast, contain their own DNA, RNA, and ribosomes. It is as though each of these organelles was at one time a free living cell that became swallowed up by the mother ship during the course of evolution.

At first, the suggestion that the nucleus, the mitochondrion, and the chloroplast had evolved from bacteria was met with disbelief. In its place thrived an alternative hypothesis—that these three organelles had each evolved directly from the ancestor, without any symbiosis.[18] But over the past half century, multiple lines of evidence have come to the support of the Serial Endosymbiotic Theory. Some of these lines arose from fine old techniques such as the microscopy of Robert Hooke, and some from new techniques such as biochemistry, and the molecular sequencing devised by Carl Woese and his colleagues. Indeed, so much unfiltered evidence has become available that there have seemingly been almost as many interpretations of the story as there are voices clamouring to be heard. Some enthusiasts saw a different sequence of the partnerships, here the nucleus first, or there the mitochondrion first. Others sought out completely different hosts, or even different guests. At times, it can seem as though all are agreed that there is going to be one hell of a good party, but none can agree who is invited or where it will take place. And a classic case of this mix-up comes with *Pelomyxa*. Now this is a story that takes us to my home town of Oxford. That is where I first heard this yarn, from Lynn Margulis when she joined me in teaching for a year. But most importantly, Oxford is the place where the infamous Elephant Pit was dug.

The Elephant Pit

Glancing up at the inside of the Oxford Natural History Museum, it is commonplace to be struck by the skeletal appearance of the building and its pillars, and by the pachyderm-like panels of its great glass roof (Plate 8). Standing beneath this roof, which dates back to 1860, it can feel as though we have crept inside the belly of a vast elephant-like creature, with all its ribs and phalanges, spines and spandrels sticking out of the ground. Indeed, as we look around this great Fossil Hall, we can even see the defleshed remains of an elephant, as well as a woolly mammoth and a mastodon, still seemingly standing in wait for dinner time, and for tending by their keepers—who sadly have never returned. It is the bones of one particular Asian elephant in this museum, though, which lead us to some vital clues concerning questions about the most primitive protozoan cells.

It is no easy matter to assemble the skeleton of an elephant—Asian or African. There is a great mass of meat that needs to be removed and disposed of. In the present case, the Oxford Museum elephant is said to have come from a local circus with which it had been travelling before it died. Always eager to add to their collections—which also included the last Dodo and the bones of the last man to be hanged at Oxford jail—the curators had the elephant delivered to the back door of the museum. Now the flesh had to be disposed of somewhere near at hand. One might like to think that this poor old elephant provided for a massive Christmas feast, allowing the dinner plates of a dozen or more Bob Cratchetts and their kin to be piled high with roast elephant, mashed potatoes, and gravy. A truly Dickensian scene indeed. Alas, the truth is darker. The curators had other plans.

Where better to do this dirty deed, thought the curators, than in a pit dug in the yard at the back of the museum? This hole in the ground was charmingly called the Elephant Pit, until it was later filled with water, when it became the Elephant Pond. It was into this pit that the elephant was pitched, and the flesh was flayed and then allowed to decay, to be stripped away by time and tempest. Left behind in the pit were those gleaming white bones, for later assembly in the museum hall.

But the Elephant Pit was to bring forth another, even darker little secret. Little by little it began to attract its own mini-ecosystem—a fuzz and froth of strange looking organisms. Forget about the woolly mammoth. And forget about the *Tyrannosaurus rex* that now graces the museum. These strange looking microbes were, and they still are, among the greatest oddities of the cellular world. They included a mammoth among single-celled creatures, a monster among monads: the amazing amoeba called *Pelomyxa pelustris*.

The blob of *Pelomyxa* ranges from as little as 0.5mm as to as much as 5mm across. That is huge in comparison with most single-celled organisms, which are usually little more than a few thousandths of a millimetre wide. Much of what we know about this strange organism has emerged from studies here, in the Elephant Pond, where it lived out its reclusive life on the muddy bottom, feeding on algae and bacteria. As might have been expected in such pond, the oxygen levels in the water could be rather low, and this seemingly suited this little blob right down to the ground.

Rather pleasingly, it was discovered by a female student, Lillian Gould, whilst working in the museum during the summer of 1893, and it was reported to the press shortly thereafter.[19] With the aid of a series of mirrors set up between ground glass lenses—called a *camera lucida* micro-scope—and a collection of tints and stains with the delightful names of eosin, carmine, and gentian violet, Lillian Gould was able to point out some of the curious internal structure shown by this monstrous monad. She was able to confirm the presence of rubbery and flexible arms called pseudopods, which serve as the food gathering apparatus. And the pres-ence of a captain—the nucleus. In fact, some generations of *Pelomyxa* can have multiple nuclei, each now known to contain the DNA for all the recipes of cooking and breeding. Further work was to show, rather strangely however, that *Pelomyxa* does not have an engine room—a mito-chondrion.[20] That seemed very odd.

How, then, does *Pelomyxa* manage to get along without an engine—a mitochondrion or two? There is no evidence that it steals, in the manner of an internal parasite. Instead, it seems to get along by taking in lodgers. Any cellular outfit can take in lodgers—endosymbionts. But there are bad lodgers and good lodgers. Bad lodgers are like ship rats, stealing the flour and leaving droppings in the pies. Only the rats are happy with this arrangement, of course. Indifferent lodgers are more like seagulls that live in the rigging but do no harm to the ship below. At the other end of the scale, good lodgers are like passengers who perform useful functions for the host, in response for a nice little place to live, and a scattering of free handouts. To call these 'lodgers' is a bit too politically-correct, per-haps. 'Servant' might be a closer fit. But it is this kind of lodger which *Pelomyxa* takes in—lodgers called methanogenic bacteria. Under low oxygen conditions, these bacteria sit in a ring around the nucleus, in effect helping the cell to metabolize, and belching out methane gas as a by-product.

This curious absence in *Pelomyxa* of an engine room—a mitochon-drion—has been interpreted in one of two ways. On the one hand, it has been regarded as evidence for the primitive nature of its outfit—like a sailing yacht that tries to get along without an engine. That was the view of Lynn Margulis and her co-workers. On the other hand, it has been regarded as evidence for the loss of an important facility—like a ship with a busted boiler. Or more properly, a boiler than has simply fizzled out owing to a lack of oxygen. That is how this loss is currently explained by most workers on this group.[21]

New walls for old

The city of Oxford also brings us within reach of another voice singing about the origins of the eukaryote cell. This line of the chorus has been sung with gusto by a microbiologist called Tom Cavalier-Smith. A key witness to the history of life, says Tom, lies in the architecture of the cell, and most especially in the nature of the cell membranes themselves. Using such evidence, plus the explosion of work upon DNA and RNA sequences, he has sought to sketch out the outlines of the great evolutionary tree. Indeed, he has planted sufficient evolutionary saplings to build a forest. But then again, he has later returned to some of these new branches for a spot of pruning. Along the way, most of these branches have been blessed with shiny new names. Some of these are euphonious, such as Unikonta and Bikonta. Several are certainly handy in conversation, such as Posibacteria and Negibacteria, or yet again Lobopoda and Brachiozoa. A few have sat in limbo awaiting their moment of glory, such as Neomura. But there are some that have me cradling my aching brain, such as Chromalveolata.

Now it has to be said that this process of evolutionary forestry followed by pruning, or even felling, has its advantages—at least some of these names, and the concepts they enshrine have proved extremely useful for describing relationships between living things. But it has its drawbacks as well—lots of trees can make for a taxonomic jungle. And a jungle is a great place for guerrilla warfare.[22] But that is maybe how it should be. This a fast evolving field. Feelings are running high, not least because the scientists have a real *passion* for working out the true course evolution of life on our planet.

Tom Cavalier-Smith sees the living world in a very distinctive way.[23] For Tom, the most ancient bacteria were not the heat-loving

Archaebacteria[24] so beloved of many early life reconstructions. Instead, they were Gram negative Eubacteria, a group which includes both *Yersinia pestis* of the Plague and oxygen-producing cyanobacteria like *Oscillatoria*. In his view, major changes in the outer cell membrane, leading towards heat-loving Archaebacteria, didn't take place until very late in Earth history. This new kind of wall is called by Cavalier-Smith 'Neomuran'— literally meaning a new kind of wall in latin. It was this Neomuran Revolution that led towards Archaebacteria, and it was from these that the eukaryote cell developed.[25]

A key viewpoint of Tom Cavalier-Smith sees the cell membrane and the nucleus both arising directly from within an Archaebacterial cell, without the symbioses implied by Lynn Margulis. This new improved Neomuran wall was thinner and simpler and proved marvellous for farming eubacterial symbionts. It was this new wall that allowed the enslavement of both the mitochondrion and the chloroplast.[26] Of these eukaryotes, as we have seen, there were two kinds: those with one flagellum, called the Unikonts, and others provided with two flagella, called the Bikonts. It was from the Unikonts that the amoebae, the fungi, and the animals arose. But it was from the Bikonts that the forams, algae, and land plants arose.

Now there are many fascinating ideas to ponder here. At one end sits the view of Lynn Margulis, who traces the nucleated cell back to a heat-loving *Thermoplasma*-like Archaeobacterium having a flexible outer cell membrane and a marked dislike for oxygen. The flagellum (or the cilia) is then traced back to a corskrew-like spirochaete bacterium. The mitochondrion arose next, from an oxygen-dealing proteobacterium. And last but not least comes the chloroplast, from a photosynthetic cyanobacterium.[27] But others such as Tom Cavalier-Smith question whether the evidence for all of this is yet good enough. They prefer to see the early stages—cell membrane, endoplasmic reticulum, microtubules, nucleus, and flagellum, all arising directly from an Archaebacterial ancestor.

There are two things that now deserve a clear emphasis. First, we can see that there is much common ground. It is broadly accepted that chloroplasts arose from the capture of cyanobacterial cells by some kind of eukaryote host, already provided with a nucleus and a mitochondrion. Second, it is widely agreed that the algae—and hence the land plants to which they gave rise—gained their chloroplasts in much the same way as those algal farming forams, such as *Archaias*. That is to say, plants and algae most likely arose from engulfment by a host having a very elastic outer membrane. An elastic outer cell membrane, indeed, very like that of a living foram. It is no surprise then, to find that the molecules inside the RNA of algae and forams show that they are sister groups. They have a shared family history.

The greening

While I was studying botany at university in the 1960s, the idea that chloroplasts were once free living bacteria would have been thought rather daft, if it was ever thought about at all. But it has since been confirmed that the ribosomes of mitochondria and chloroplasts resemble those found in bacteria, not those of the host eukaryote cell. As if to emphasize this fact, they even fall sick when treated with bacterial antibiotics, such as Tetracycline.[28] Not only that, but both chloroplasts and cyanobacteria have similar molecular pathways for photosynthesis, consistent with a shared ancestry. Most important of all, chloroplasts, like the mitochondria, have their own versions of the double helix of DNA. Their DNA is circular like that of bacteria; it contains genes not to be found in the nucleus of the same cell; and this distinctive DNA also reveals a genealogy for all chloroplasts that is quite different from that of the nuclei.[29]

It is therefore hard to avoid the conclusion that chloroplasts were once free living cyanobacteria that became enticed to live inside the cytoplasm of some kind of host. This relationship may have started out like that between forams and their symbionts in the Caribbean reefs—with benefits to both sides under nutrient deficient conditions. But there is a huge and awkward distinction that we now need to make between the foram-algal symbioses of the Caribbean reefs and the chloroplasts found within the cells of nearby algae and plants. This distinction relates to the problem of the DNA heist.

The DNA heist

Symbiotic algae living within forams and corals still have their full complement of DNA and RNA. That means they can live independent lives if they so wish to do, as happens when the consortium encounters a Green Water ecosystem. But chloroplasts found within those very same algal cells do not have this option. They cannot live free and independent lives, whatever the conditions might be. How has this contrast come about?

Over the vastness of geological time, it seems that algal chloroplasts have been made to render more and more of their personal DNA unto the host cell, where it has been gradually incorporated into the nuclear DNA of the host. In most cases, little more than 10 per cent of the original complement of DNA now remains within either the chloroplast, or indeed the mitochondrion. Some of this DNA has been lost, presumably because neither chloroplasts nor mitochondria were required to make all their own proteins once they were being comfortably fed by their hosts. But this meant they were increasingly subjected to control by their host's own DNA, and were no longer free to escape.[30]

It is not at all clear how this transfer of genetic material took place from guest to host. That is because DNA—on which the genetic material resides—is usually thought to be too large to pass through the kinds of complex membrane that surround both the host and its symbiont. But there is some evidence which may help us here—from the biology of viruses.

A virus does not possess a cell membrane of any kind. That is why a virus is not considered to be either a cell, or indeed a free living organism. Instead, it is entirely parasitic on other cells. For this parasitic way of life, a virus does need a genetic code, in the form of DNA or RNA. But it also needs an ability to penetrate the cell membranes of its host. Indeed, it performs this act of penetration with great aplomb, as we know from painful throats during a bout of 'flu. Of considerable interest here, then, is the capacity of viruses to transfer genetic information from one eukaryote to another during an attack. This horizontal gene transfer has been used, for example, to explain why the genes of a wild gerbil actually carry splices of snake genes.[31] Such processes of horizontal gene transfer are also known to be widespread between bacteria, as shown by the molecular studies of Ford Doolittle of Nova Scotia. Indeed an orgy of gene exchange seems to have taken place in the past, which has caused a massive hangover for those involved in molecular studies.[32] It is possible, therefore, that prolonged exposure to viral or bacterial pathogens could have assisted the horizontal transfer of genes from guest to host.[33] Possible, but as yet uncertain.

Figure 23. Lynn Margulis (top right) in symbiosis with her friends at a meeting in Berlin during July 2011. Also shown are the bacteriologist Betsey Dyer (far left) plus scientific translator Zoë Brasier (bottom left) and the author (bottom right).

Whatever the mechanism, it is known that the host cell releases helpful porters[34] that help to ferry useful molecules across the chloroplast membrane, and to help to keep it happy. And that the chloroplast produces photosynthetic gifts such as sugars that help to keep the host happy in return. Both seem to be winners here.

Nested Russian dolls

There now follows a troubling tale. It is a story of successive enslavements. This harrowing tale has been reconstructed from the mapping of both cell membranes and molecules. Here is how this saga is said to run.

Green algae, like those of *Enteromorpha* that we met in the dockyard, have chloroplasts which are surrounded by a double walled membrane. Exactly the same is true for a Red alga like *Lithophyllum* that clings to the coral rubble on the reef crest. In both cases, the outer layer is taken to be that of the host cell.[35] And the inner layer is thought to be the remains of the original cell membrane of the cyanobacterial symbiont. One ancestor of this kind may have resembled living *Prochloron*, a cyanobacterium which has a mix of photosynthetic pigments very like those found in eukaryotic Green algae. And another may have resembled living *Synechococcus*, a cyanobacterium with pigments closer to those of Red algae.[36]

In this way, we can account for the origins of not only the Green algae and the Red algae, but also the rare and mysterious glaucophytes found in freshwater lakes. Each of these had chloroplasts that arose from a single symbiosis.[37] But many kinds of algae have been found to display not just two-ply walls of this kind, but even three- or four-ply walls around their chloroplasts. This paradox makes sense, however, if those extra walls were the result of secondary or even tertiary symbiotic events.

Thus one host protozoan is believed to have captured not a cyanobacterium but a Green algal symbiont, just like we have seen inside the reef dwelling foram *Archaias*. This capture is revealed by the presence of a three- or four-ply membrane around the chloroplasts. Such a partnership resulted in the evolution of a group of multicellulae algae called the chlorarachniophytes—a sister group to the forams—as well as another group of mainly single-celled forms called the euglenophytes.[38] But yet another group of related organisms called the trypanosomes seem to have lost these chloroplasts altogether and become dangerous parasites. It was these *algae-minus-chloroplasts* that likely infected Charles Darwin in Chile, to give him a lifetime of misery with Chagas disease.

Yet another protozoan is inferred to have captured a Red algal symbiont, just like we find going on inside the living reef dwelling foram *Peneroplis*.[39] In this case, the partnership resulted in the evolution of Yellow Green xanthophyte algae, as well as the Brown phaeophyte algae, and also the lacy diatoms, and the golden coloured crysophytes.[40] In

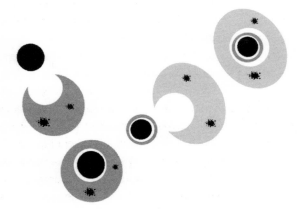

Figure 24. Nested Russian dolls. A succession of symbioses took place during the early evolution of the algae. These multiple liaisons are inferred from multiple walls that line the chloroplasts. From left to right in this cartoon, a cyanobacterial cell (black) is engulfed by an early eukaryote cell (dark grey) to produce an alga with chloroplasts having a double membrane (at bottom left). Both Red algae and Green algae were the result of this enslavement. Such algae were then engulfed by a different suite of eukaryotes (light grey, at centre) to produce groups such as the Brown algae having a triple membrane (at top right). Such a secondary, and even a tertiary, symbiosis arguably happened several times over, to produce the different algal groups we see today. At each step, the internal symbiont lost crucial genes to its host, making escape impossible.

some of these, the chloroplast was again lost, to give rise to that fungus-like group of pathogens which cause Dutch Elm Disease.[41]

But, for some modern researchers, this is not an end to the matter. A curious case has been argued for cells of certain dinoflagellates, allied to the symbiont *Symbiodinium* we found living inside reef corals. These have been argued to show evidence for a third kind of marriage.[42] In this case, we must infer that the ancestral protozoan harboured a Brown algal symbiont, in the manner of those reef dwelling forams we have met with, called *Borelis*, *Amphistegina*, and *Heterostegina*.[43]

If this saga is true, then there was not just one symbiotic heist but up to half a dozen. But for some, including Tom Cavalier-Smith, some of this sounds a stretch too far.[44] Whatever the outcome, we have good reasons to suspect a succession of symbiotic partnerships leading towards enslavement (Figure 24). And for multiple cross branches in the tree of life. Or rather, the Mangrove Tree of Life.

Only connect

We are accustomed, perhaps, to think of progress in science, and the progress of society too, as a steady and gradual process. As the years go by, the ideas steadily fall to the ground like a blanket of leaves. We then imagine this layer growing steadily thicker, as our human understanding rises upwards towards the light. But looked at more closely, something much more interesting is going on. It is true that the information keeps falling like leaves to the earth. But the data are then gathered up by strong vortices of wind—currents of thought—which pile on the ground in heaps that can remain stable for years. Leaves may fall in one place but they soon get swept along to another place, where they begin to settle. These currents in science—called paradigms by Thomas Kuhn—can remain intact for decades. Every now and then, though, the wind pattern changes. A huge gust comes along, and the leaves start to accumulate in another part of the garden entirely—a paradigm shift takes place. It would be foolish to think that these shifts mean that science never makes any real progress, that it just shuffles the data around. This is what a few postmodernist historians have argued. But one only has to look at the ever rising thickness of the pile of leaves to see the shallowness of that hypothesis. The layer of leaves—human knowledge—continues to grow higher and higher towards the light, whatever the paradigm. But this layer is seldom flat. The leaves accumulate in piles.

It is in this way that we might begin to view these shifts in thinking about the evolution of the cell. One the one hand we have one big pile of leaves—a paradigm—which gives weight to the importance of the individual and the hierarchy. But leaves of thought have continued to accumulate on the ground until eventually there have been enough to form another pile, swept up by a different current—the importance of networks in evolution, down to its deepest levels.[45] We can learn something from this about the way in which science works in our world. Science, remember, *is* one of the things that a democracy *does*. As with any democratic system, it happily embraces doubt. But science goes one step further. It takes this doubt for its banner. It is hugely concerned with questions about the nature of the universe, and the measurement of doubt. Once predictable answers to its questions have come to hand, then these are allowed to slide comfortably into bed with technology and culture.

Winds have swirled and debates have stormed about the relative importance of these two concepts—of individuals and networks—during our lifetimes. But the importance of understanding networks is now sharply on the rise, as we shall shortly discover.

The crunch

So what are we to make of all these symbioses? Might there be something rather fishy going on here? Well, indeed, there might.

In the early chapters, remember, we came across evidence that living reefs around the Caribbean have tended to encourage symbioses within corals and Foraminifera. Low levels of free fertilizer, such as nitrate and phosphate—Blue Water conditions—have encouraged cells of these creatures to take in lodgers, such as those of the Green, Red, or Brown algae. Together, they can result in a world turned outside-in—the algae have moved inside their host cells. This process of taking in and cultivating algal symbionts, we may here call *symbiotectis*—the building of symbiotic structures.[46] We can easily observe this kind of thing not only in reefs themselves but also in ponds among protozoans like *Paramecium* with its Green algal symbionts, and even in lichens with their consortia of fungi and Green algae.

But later, we have met with worrying evidence showing that such symbioses can be turned completely inside-out again, especially if the waters get too warm, or fertilizers get too rich—Green Water conditions. That is what seems to be happening in places today, where symbiotic algae are being expelled from their host cells. Coral bleaching and Black Band disease—an infection by nutrient-loving cyanobacterium *Oscillatoria*—are two kinds of symptom that show this decay. Such a process of destruction we may here call *symbioclasis*.[47]

Together, these two phenomena are like the two sides of a silver dollar. Heads up: the symbioses are allowed to build up. Tails up: they can be made to collapse again. This flipping and flopping is, therefore, what we might expect from stories of the living world. And from pet shops too.

But what Lynn Margulis and her colleagues have told us goes far beyond all this. At some unknown time in the long distant past, a protozoan-like host and its algal symbionts were actually able to sink all their petty little differences and become fused together into a wholly new class of organism.[48] Yes, there were flips of the coin. But somehow, no flops. The flips of chance just kept on producing ever stronger and stronger symbioses.

In some curious way, it seems that the ancestral hosts of our modern cells got very lucky indeed. They hit a real winning streak. And this didn't just happen once. It happened several times over. First flip. Heads up, and the nucleus formed. Second flip—heads up again! And the mitochondrion formed. Third flip—heads up again! And the earliest green or red chloroplasts were formed, by enslaving cyanobacterial cells. Fourth flip—heads AGAIN! And triple walled, brown chloroplasts were formed by capturing Green or Red algal cells.[49]

Now, that sounds like a very lucky streak indeed. But as far as we can tell, nothing quite like this is still going on today. Only in the past has this happened, and only with the simplest organisms, too.

We ought therefore to make an important distinction here. Those symbioses between corals, or forams, and their symbiotic partners are akin to a modern civil partnership—a contract has been signed between the partners, but the passports and entitlements of both partners have still been kept apart in separate bank vaults. But something a bit more radical has come about, though, during the origins of the chloroplast and the mitochondrion. Both relationships are more like enslavement, in which the senior partner has locked away all the legal documents—diners club card, passport, and even marriage rights—of the junior partner. More mordent still, this slavery has become permanent. It has been handed down from one generation to the next generation without any hope of freedom.

This long succession of lucky strikes for the host cell went much further than is now seen in modern coral reefs—the process of symbiosis building, or symbiotectis. No matter how much Nature might chose to torture a nucleated cell or a Green algal cell, the cell in question will always 'refuse' to expel its organelles. That is because those organelles are completely unable to swim away on a whim to lead free and independent lives as they can in corals or forams during bleaching episodes. Organelles—including our icon the chloroplast—are instead bonded together with their hosts 'till death do them part'. This is not symbiotectis but *symbiogenesis*, and it is very rare and strange indeed.

So when did this run of symbiogeneses actually take place? And under what conditions did the change from partnership to total enslavement of those chloroplasts happen? To answer this, we will need to dig deep; very deep indeed, far down inside the rocks of the fossil record. Happily, there is no better place to begin this quest than along the banks of the Nile.

Time, at long last, to pay our respects to the Sphinx.

CHAPTER 8

RIDDLE OF THE
SPHINX WITHIN

Strabo's lentil

One day in 1982, and near to dawn, I found myself standing along the edge of the Giza plateau—staring up at the Sphinx. There are few sights more stirring than the Sphinx at sunrise. Although small by comparison with the Great Pyramid that looms behind, its great craggy head, with defaced features and mysteriously bland expression connected with something deep inside. What surprised me even more, though, was the fact that the Sphinx sits inside a kind of quarry, carved out of the bedrock to make space for the chiselling of its cat-like claws (Plate 9). On that winter's morning, I had rather reluctantly agreed to have my photograph taken beside this curious hybrid creature. And as I stepped into the gully I noticed, perhaps from force of habit, the creamy brown limestone that lay all around, with metre-thick bands etched out by millennia of wind and rain. Even the flanks of the Sphinx looked worn and banded, so that the creature seemed to morph out of the very rock itself. But as I moved away, my eye was caught by a curious object lying near the quarry floor. It was an example of Strabo's Lentil—an object with its own monumental story to tell.

Strabo's lentil first came to notice around 20 BC, when a delegation was passing through the newly established Roman province of Aegypta. Among the curious travellers rode a young Greek called Strabo, perhaps acting as a spy for the emperor Augustus.[1]

I like to imagine the scene of Strabo's revelation as having taken place on a similar winter's morning,[2] when the air above the desert often fills with yellow dust. Whilst shaking the sand from his ears, Strabo caught sight of something rather strange. Perhaps he let out a little gasp of surprise. Maybe he just kept calm. But beneath the aprons of dust, Strabo evidently saw mysterious shapes, seemingly locked *within* the cold framework of the unforgiving stone. Bending down, he could see that such patterns were lying in the stones all around him. Years later, on recollecting his thoughts and checking his notes back home, Strabo the Greek went on to recall his moment of revelation like this:

> One of the most marvellous things I ever saw—near the pyramids of Egypt—consisted of heaps of stone chips; and among these heaps I was able to find flakes that resembled lentils, both in form and size. Not only that, but under some of these heaps could be found the 'winnowings' of things that look rather like half-peeled lentil grains. My Egyptian guide explained to me that these 'lentils' are thought to be little scraps of food, left behind long ago by the pyramid builders, which later turned to stone. That is not an altogether improbable explanation.[3]

Now, we know from his other writings that Strabo had also seen sea shells set far inland, near his home in Asia, and had pondered what these might mean for the history of the planet. But he had never seen anything quite like these markings before. Silly stories about stones were the stuff of legend back then, and, even today, they provide 'the hook' for a gullible public. Strabo was seemingly determined not to make up a series of silly stories about stones. His old teacher, a true Stoic, had once told him 'never be amazed, and never marvel'. Strabo therefore interpreted these structures as lentils—dried peas—that had become petrified after they fell from slave's lunch packs.

His hypothesis of fossilized lentils was a sensible one for its time. But Strabo's 'lentils' were actually more amazing than fossilized lunch packs—indeed, they were odder than anyone would have dared to think. That is because Strabo had accidentally stumbled upon the real Riddle of the Sphinx—an improbable mountain of fossils, all set in stone around the monuments of the Great Pyramid of Egypt. These were the first fossils ever interpreted as such in human history.

We now call these fossils *Nummulites*—meaning coin stones.[4] Indeed, the Pyramids and the Sphinx are riddled with them. This odd-looking layer can then be followed across the Nile to the hills of Mokkatam on the east bank, whence further blocks of stone have been quarried for building. But there are even stranger things to relate. This same bedrock with its fossilized *Nummulites*, give or take a few million years, has been uplifted from the sea to form a rib of rock that winds across half the planet (Plate 10a). From Selsey Bill in southern England, it sets off southwards towards the Atlantic coast near to Biarritz in France and then wends its way through Italy, Greece, Egypt, and Asia Minor towards the Zagros Mountains of Iran, and thence towards Oman along the Arabian Gulf at Muscat. From there, it crosses over the Arabian Gulf and heads off into the embattled mountains of Afghanistan. Once there, it tilts skywards and marches towards the rooftop of the world, in the Himalayas of India and Pakistan. It can be spotted near the end of this great journey, dropping thankfully down into the warm Pacific Ocean near Thailand.[5]

From this we can learn that *Nummulites* was at one time thriving across a vast tropical seaway that no longer exists, called the Tethys Ocean. It no longer exists because its hinterlands have long since squeezed that ocean out of existence, with Africa bumping into Europe to form the Alps; Arabia slamming into Asia to form the Zagros Mountains; and India smashing into Asia to form the Himalayas. Those mountains still exist today, of course, but in the process of their formation, the ocean itself has largely disappeared.[6] And nothing exactly like these fossils still survives either. For these are the remains of organisms that once lived along the flanks of the Tethys Ocean during one of those time periods that were named by Charles Lyell—the Eocene, some 50 million years back—which was also the warmest period during the last 100 million years or so.

During the Eocene, reef-building corals were supposedly finding it a little too warm.[7] But other forms of marine life were flourishing. Among these were the earliest whales, evolving rapidly from some kind of hippopotamus-like ancestor. Nor were whales the only giants to emerge during this warm period. Giant Foraminifera like *Nummulites* were also starting to flourish.

Wherever they occur, these *Nummulites* laid down beds of limestone that now look a bit like fossilized lentil soup. When we look closely at the architecture of their shells, we can spot features that look vaguely familiar from our travels across the reefs of the Caribbean. Their skeletons recall again the architecture of the Colosseum of ancient Rome, replete with its chambers and cubby holes, doorways and secret passages. More especially, they call to mind a simplified version of *Heterostegina*—the protozoan we met around the reefs of Pedro Bank; or *Amphistegina* that we met in the caves of Barbuda. And like those two, we can assemble a shopping list of features which indicate that *Nummulites* was a keen cultivator of symbiotic algae.[8]

Protozoan dinosaurs

The protozoan remains that make up the body of the Sphinx and the Great Pyramid can be huge—up to 40mm across. But *Nummulites gizehensis* is not the largest of these. Even bigger ones have been found in rocks of this age, called *Nummulites millecaput*. These monsters can reach up to 100mm across.[9] Puzzled by this gigantism, I went to my departmental library to get hold of a book on fossil *Nummulites*.[10] There was a surprising amount of information available about these monster monads and their fossil relatives—hundreds of species have been described ranging from weedy little shells the size of a poppy seed to giant ones the size of a pancake. Such fossilized *Nummulites* are real attention-seekers, and they have been bugging geologists for well over a century. That is because the limestones in which they occur are of fundamental importance to the economy of the world. Let us take a closer look at this for a moment.

At the time when these protozoan dinosaurs were flourishing on Earth, some 60 to 40 million years ago, the main reef builders were not corals. They seem to have been *Nummulites* and their relatives. The reason for that switch in dominance is still unclear, but it may have had something to do with large amounts of greenhouse gas in the atmosphere, and the correspondingly greater acidity of the seawater. This was, after all, one of the most hot and sweaty periods we know about in the last 100 million years. But there was another sticky problem, too. For whatever reason, these *Nummulites* flourished in their billions. Indeed, in their billions of billions. But they did not form cathedral-like constructions reaching up to the sunlight as do modern reefs. Instead, they built huge banks, rather like dunes in the Sahara desert today, that spread across the shallow tropical seas of the old world, from England to the China Sea. The clarity of the water above these banks was very great—we can be sure of that because there is almost no mud in these fossil sand banks, hence the limestones in which they occur tend to be wonderfully pure. So we can visualize the habitat of these protozoan monsters and their living sand dunes as being true Blue Water ecosystems that girdled much of the tropics. These conditions even extended into the Caribbean too, of course, but *Nummulites* found it difficult to swim across the ever widening Atlantic, and different forms of giant protozoan are known to have flourished in their place.

A careful look at a single denizen of these sea dunes—a giant *Nummulites*—shows that it must have been cultivating algal symbionts within its tissues, much as does *Heterostegina* today. We can deduce this not only from its Blue Water habitat—indicating that phytoplankton was scarce—but also from its skeletal architecture.[11] Not only was *Nummulites* constructed in the manner of a Colosseum, with myriads of tiny little chamberlets, each interconnected by short and efficient lines of communication, but this protozoan built a shell provided with walls of thin and translucent calcite. In other words, it had walls like the windows of a seaside solarium. These windows were able to let in the sunlight to allow for photosynthesis by symbionts sitting within the shell itself. Some species of this extraordinary protozoan even went one better, by secreting rods of calcite that seemingly acted like fibre-optic lenses, conducting high levels of light into the innermost recesses of the shell.[12] These and other features suggest that *Nummulites* was a highly efficient farmer of photosynthetic symbionts, most probably of Golden Brown diatoms like those cultivated by its living relative *Heterostegina*.

In this way, great banks of *Nummulites* limestone were being built up above the seafloor. Being made of limestone—chalky calcium carbonate—these rocks are always full of cracks and holes. And being full of gaps, they are able to play host to some very useful fluids—ones that like to sneak through the rocks at depth. The first of these liquids is potable drinking water, so that *Nummulites* limestones form excellent aquifers in places like the Sahara and Arabian deserts where water for drinking and crops is much in need. But the second sneaky fluid that lurks in the depths of the Earth is that black gold called petroleum—the hydrocarbon residues left by the decay of phytoplankton that settled on the seabed long ago. No surprise, then, that the secret service of the oil industry[13] has taken to mapping these limestones with some precision at depth, using a mixture of geophysical profiles, ground-truthed by drill cores.

At this point it has to be admitted that the vast majority of working palaeontologists have not been chasing fossilized lizards. Nor have they been trying to solve climate change. They have been trying to plot those changes in fossils that can be read from the rocks, as each drill bit chomps deeper and deeper into the rocks in search of oil. By the 1920s, Joseph Cushman of Massachusetts, among others, had discovered that a similar story of evolution and environmental change can be coaxed from the rocks as one descends backwards through time. He showed that few fossil groups are more useful for this work than are the shells of Foraminifera. That is because they are small and abundant, as well as complex and diverse.[14]

Hundreds—possibly thousands—of people have therefore toiled at identifying fossil species of *Nummulites* and their relatives from chips of core or drill bit from oil wells around the world. These and other microfossils have thereby helped teams of geologists to formalize a timeframe for the history of the planet, and for its environmental evolution too. And this is a history we now need to explore.

Eozoon

Nummulites has a curious place in the history of life. For a quarter of a century or so, this fossil was believed to be a relative of the most ancient organisms on the planet. The trouble started when a very senior Canadian geologist, Sir William Logan, found the remains of things that looked a bit like *Nummulites*, but in vastly older rocks. These were not a mere 50 million years old, like the rocks of the Sphinx; Logan's rocks were several billion years older, from the so-called Laurentian rocks near Ottawa in Canada. Later called *Eozoon*, this outrageously stripey marking convinced Dr William Dawson of Montreal, to write peons of praise about the presence of Nummulitic Foraminifera in some of the oldest known rocks.[15] That might not have mattered much. But the world's most eminent microscopist, Dr William Carpenter, then put his full weight behind the claim.[16] That might not have mattered much either, but the world's most eminent geologist, Sir Charles Lyell, then threw his name behind the claim.[17] That might not have mattered much, either, but Mr Charles Darwin then threw his name behind the claim, by entering it as evidence for early protozoan life in the later editions of his *Origin of Species*.[18]

The English establishment had dug itself in deep, but things did not turn out well for them. Piece by piece the evidence started to roll in. *Eozoon* was not an early example of *Nummulites*, or indeed anything biological at all: it was a product of the cooking and squeezing of the rocks under high pressure; a pattern produced by metamorphism. Neither Carpenter nor Dawson could find any wiggle room but they refused to relent. Both therefore went to their graves as deeply troubled men.[19]

Nummulosphere

As soon as the world had disposed of one myth about *Nummulites*, it was obliged to confront yet another. The eccentric world of this monster protozoan was to be taken to further extremes by an Edwardian scientist at the Natural History Museum in London. The writings of Dr R. Kirkpatrick were published during World War I in a series of rare volumes that, luckily perhaps, escaped wider notice during the madness of battle.[20] His theory was simple and direct—that the whole of planet Earth is made entirely out of *Nummulites* shells. This view he extended not just to limestones and chalks, but to coal, to volcanic lavas, granites, diamonds, meteorites, the rust at the bottom of his saucepan, and even to the dung of own his pet dog. This is what he tells us in his own words: 'Volcanoes and lava streams are heaps and masses of silicated nummulitic limestone; i.e., they are of aqueous and organic origin. All the rest of the igneous rocks are of a similar nature and origin. Seeing that these rocks and the sediments derived from them constitute the bulk of the planetary crust, it follows that the lithosphere is mainly composed of silicated nummulitic rock.'[21]

This bizarre thesis of 'the Nummulosphere' was argued with surprising attention to detail, assisted by ink-drawn sketches or sepia photographs taken down his own microscope. But old Kirkpatrick was sadly deluded about his fossils, of course. We can easily see that the whole world is not made up from *Nummulites* shells. They are only to be found in sedimentary rocks from around the Eocene period; only from tropical settings; only from rocks laid down in the photic zone; and only from rocks of the Old World. While some protozoans were surely to be found in the dung of his beloved pooch, they will not have been nummulitic protozoans.

We can now see that both Dawson and Kirkpatrick had fallen for that old Mofaotyof ('My Oldest Fossils Are Older Than Your Oldest Fossils') trick—seeing what we expect to see, and not what is really there.[22] While we can laugh at these ancient follies, we need to remain careful too. Cases of mistaken identity are easily made, but they are usually kept concealed. Every scientist worth their salt keeps a fine case of Mistaken Identity stored at the back of his or her metaphorical wine cellar. And we also need to remember this: that big questions = big mistakes. Any scientist who wishes to avoid making big mistakes is therefore doomed to fuss over little questions. Many therefore choose to do just that, not least because there are ample rewards in the measurement of little things— respectability, social rewards, medals, and grants. Old Dawson, and then old Kirkpatrick, were not of that ilk though, proclaiming their monstrous theses to the world, and allowing us to learn from their mistakes by example.

It is worth reminding ourselves again that we are never told what the name of the game is with the fossil record. Armed with the advantages of hindsight, we can see that both Dawson and Kirkpatrick had not only guessed at the wrong answer. Both had somehow guessed at the wrong question. Their wrong question was this one: what do these objects remind us of? A more proper question should have been: *what are these objects in reality?* It is a very common mistake indeed in palaeontology.[23]

A sorry tail

But there is a genuine puzzle about giant protozoans such as *Nummulites*— and their symbionts—that we now need to address. If forams and their internal symbionts have been so devoted to each other, why didn't they tie the knot, and combine to form a new kind of organism—or at least a new organelle—given the vast expanse of geological time? Such marriages, we might think, ought to be extremely common. As we have seen before, Lynn Margulis and her Serial Endosymbiotic Theory for the origin of the eukaryote cell urges us to seek them out in the living world. However, careful inspection of the living world shows that such extramarital flings are common but that firm contracts of marriage between cells have seldom been solemnized. Indeed, they seem to have been extremely rare in Earth history. It may therefore be helpful for us to look at the fossil record to see why this might be so.

Now *Nummulites* was not alone in being a major symbiont user in its day. The fossil record of foram groups that likely cultivated symbionts is rather splendid.[24] There were over a dozen different associations in the Eocene, each seemingly adapted to different light levels, to different water depths and to different types of symbiont in the reefs. One type, now called *Dictyconus*, seems to have lived in the tidal zone and possibly cultivated cyanobacterial symbionts. A second kind, called *Orbitolites*, lived in well-lit sea grass meadows, and plausibly cultivated Green algal symbionts. They flourished in the tropical lagoons around Paris that have now turned to rock. A third type, called *Alveolina*, reached rock-forming proportions[25] and looks likely to have favoured dinoflagellate symbionts. These flourished around the tropical lagoons of Selsey in Sussex, and have likewise turned to stone. And last but not least, a form called *Assilina* seemingly encouraged an internal flora of diatom symbionts, so that it was able to thrive in the deeper waters of the tropics. All over the Eocene world, it seems, little things were taking in lodgers and turning their world *outside-in*.

So successful were these protozoan colossi in the past that they accumulated as vast piles of sediment upon the seafloor. Even today, if we pick up some sand along the coast of Barbuda or along the Great Barrier Reef, it will likely be filled with the skeletons of *Archaias* or its relatives. Travelling back in time, some 50 million years or so, comparable forms such as *Nummulites* built up the limestone banks from which the Sphinx and the Great Pyramid were sculpted by ancient Egyptians. But there is more. Much more.

A further 200 million years back, into the Permian period, and we find similar forams—called Fusulinids—likewise flourishing across the tropics (Plate 10b).[26] Today, their fossilized constructions are helping to hold aloft the mighty Himalayan mountain peaks. Some of their foram shells were in excess of 150mm across. That may seem a vast size for a single-celled organism but, as we have already said, these colossi were not really single-celled creatures at all; they were filled with thousands of cells of one kind or another—various kinds of algal symbiont such as we have seen inside forams around the Pedro Bank. We can be reasonably sure of the presence of algal guests in these long dead fossils from the shape and construction of the shells themselves—which were specially provided with clear windows—and from their Colosseum-like architecture. Other kinds of invertebrate may also have been symbioholics at this, including some coral-shaped brachiopods.[27]

The extinction factor

These symbiotic liaisons hold interest for our study of the deep history of life for one rather obvious reason. They provide clues as to how the greatest divide, and hence the greatest obstacle, in the whole of the living world was crossed—that between bacteria and all higher organisms. And most especially, how and why the chloroplasts of algae and plants may have become permanently enslaved.

Today, there is no doubt that the Foraminiferan host and its algal guests are entirely separate organisms—they have very different DNA codes and, while they may synchronize their cell divisions, the algae can still swim away to live perfectly independent lives—leaving behind the poor host to starve to death. But the 'symbiotic knot' is thought to have been catastrophically untied or cut at those times we call 'mass extinctions' when unusually large numbers of creatures died out rather rapidly, such as those at the end of the Permian and the Cretaceous periods, leading to the likely death of the host and its genes.[28] It is possible, of course, that some of the algal guests could have survived these calamities, but this is something we can only speculate about because they are never preserved directly in the fossil record. But now comes a fascinating puzzle. These extinctions of symbiotic forms seem to have taken place surprisingly regularly through the fossil record—every 20 to 30 million years or so.[29] And these extinctions are not seen in the less complex kinds that we have met—temples and tombs such as those shown in Figure 15. Hence we find that *Saccammina, Ammodiscus,* and their relatives have sailed through these calamities unextinguished, from as far back as the Cambrian period. Those banana-shaped constructions of *Quinqueloculina* have nonchalantly carried on their business since the Jurassic. These are just a few examples. There are hundreds more that point up this strange contrast.[30] Colosseums collapse. But Temples survive. Such a distinct and regular pattern of collapse and survival demands a decent explanation (Figure 25).

We now come to a point where many of the stories regaled in previous chapters start to come together, to help characterize these cycles of reef building and collapse. By 1988, a possible narrative was developing in my mind as follows (noting in brackets those portions of the story already covered in previous chapters): Blue Water ecosystems (Chapters 3 and 4, as seen in the Sargasso Sea and the Caribbean reefs) encouraged intimate symbioses between corals or protozoans with elastic cell walls, and photosynthetic algae (Chapter 5). These partnerships may have owed much to the lack of essential nutrients in the waters (Chapter 5). Rapid changes towards Green Water ecosystems, and other kinds of change, may have then have stirred up the waters, meaning that the symbiotic partnerships came under stress (Chapters 5 and 6). On geological times scales, such symbiotic associations became severely stressed and prone to collapse (as discussed in this chapter).[31]

The regularity in this construction and then destruction of symbiotic associations can seem like the swing of a vast cosmic pendulum. So it may be no surprise for us to learn that it has been connected ultimately,

PLATE 1. The oak-clad seventeenth century library of St Edmund Hall in Oxford, showing the writer studying an original copy of Robert Hooke's *Micrographia*

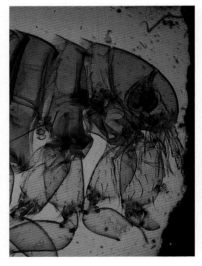

PLATE 2A. An eighteenth century flea, of the kind seen down Hooke's microscope. About 2 mm long

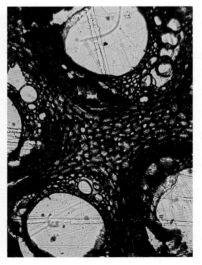

Plate 2B. Eighteenth-century cork cells, much as seen by Hooke. Field of view about 1 mm across

PLATE 3. Foraminifera abound in the coral reefs of the world. These examples thrived on the floor of a Pacific atoll during the time of Darwin's researches. Field of view about 2mm across

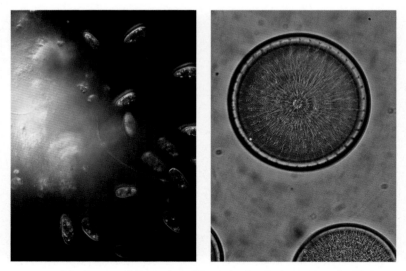

PLATE 4A, *left*. Swarms of slipper-shaped *Paramecium* bustle about using cilia. Their activities contrast with diatoms like *Actinocyclus* (PLATE 4B, *right*) which lack cilia or flagella. Field of view about 0. 2 mm across for both

PLATE 5. The lost paradise of South West Cay on Pedro Bank. Algae and corals form dark patches near the shore. My colleague Peter Dolan can be seen collecting samples

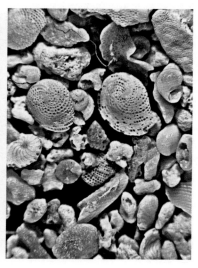

PLATE 6A. A haul of green algae, corals and molluscs collected with a snorkel near the reef shore

PLATE 6B. Microscopic shells of the foram *Archaias*, plus mollusc shells, algae and coral fragments, make up these reef sands. Field of view about 10mm across

PLATE 7. Single-celled *Discospirina* is unique among living cells. This foram recaptures 542 million years of evolutionary change during growth. The shell is about 8mm across

PLATE 8. Strange amoebae were discovered by Lillian Gould following preparation of this Victorian pachyderm in the 'Elephant Pit'. Behind can be seen the gallery where evolution was debated in public for the first time, in the Oxford University Natural History Museum of 1860

PLATE 9. Riddle of the Sphinx Within. Half cat, half man, the Sphinx provides an archetypal chimaera. Ironically, the monument is carved from the shelly remains of 50 million year old cell chimaeras called *Nummulites*, also the first fossils ever described

PLATE 10A. Slices through the Sphinx limestone reveal myriads of chambered *Nummulites* shells. These hugely successful foram-algal symbioses died out towards the end of the Eocene some 35 million years ago

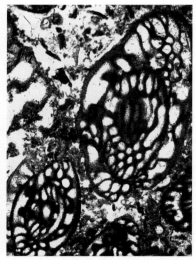

PLATE 10B. Comparable chambered symbioses, now called fusulinids, once thrived some 260 million years ago. They collapsed completely at the end of the Permian. Field of view about 2mm across for both

PLATE 11. This remote shoreline near to Schreiber, along the north coast of Lake Superior in Ontario, Canada, allowed discovery by Stan Tyler and Elso Barghoorn of the marvellous Gunflint chert microfossils, nearly 2 billion years old

PLATE 12A. The old Gunflint shoreline was made from weathered lava, seen as greenish rocks in upper left. These were then draped with grey and white stromatolitic chert

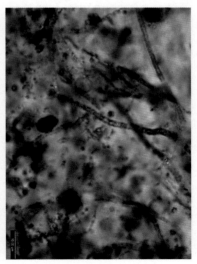

PLATE 12A. This chert is filled with rounded blobs of *Huroniospora*, and filamentous chains of *Gunflintia* cells. Field of view is about 0.08mm across

PLATE 13. The Lost City, sculptured by eons of erosion from 1450 million year old sandstones of the Roper Group in Queensland, Australia

PLATE 14A. Layers of 1600 million year old Amelia dolomite in Queensland contain no clear evidence for eukaryote fossils but good evidence for bacterial cells

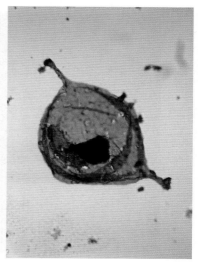

PLATE 14B. *Tappania* from the Roper Group provides some of the earliest evidence for eukaryote cells akin to green algae. Image courtesy of Emmanuelle Javaux. Fossil is about 0.1mm wide

PLATE 15. This view in northwest Scotland spans the infamous Boring Billion. The crags along the coast are Lewisian rocks some 2 billion years old. They are here draped by 1 billion year old Torridonian red beds, seen in the foreshore, and sculpted into distant hills. These bear the oldest signs of complex life on land

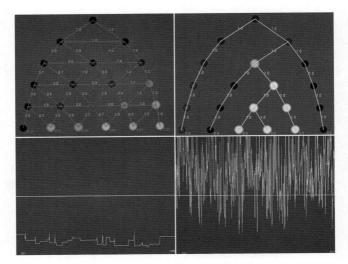

PLATE 16. Models by Matthew Brasier show how interlinked networks (at left) reveal more stable behaviour than hierarchical arrangements (at right). Small numbers refer to numerical filters at each node. Grey tones track numerical shocks moving through the system. Such models suggest that hierarchical systems do not collapse from the size of the shock but from the patterns of interconnection

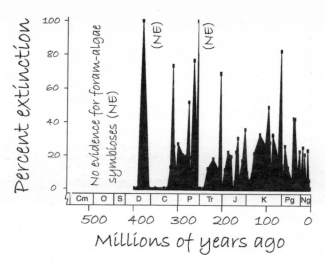

Figure 25. The riddle of the Sphinx Within. This graph charts the numerous collapses (black areas) of foram–algal symbioses, beginning in the Devonian at left and continuing right up to the present day at right. This means that the 'Sphinx Within' has developed and then collapsed over a dozen times through the latter part of Earth history. Observe the three episodes with no evidence (NE) for such symbioses.

Note. Collated with the help of Jonathan Antcliffe, and derived from many resources including Brasier (1986) and BouDhager Fadel (2008). The scale along the bottom is as follows: (Cm) Cambrian, (O) Ordovician, (S) Silurian, (D) Devonian, (C) Carboniferous, (P) Permian, (T) Triassic, (J) Jurassic, (K) Cretaceous, (Pg) Paleogene including Paleocene, Eocene Oligocene, (Ng) Neogene including Miocene, Pliocene, Pleistocene, and Holocene.

by some scientists, with all those multifarious causes invoked for the death of the dinosaurs—cosmic radiation events, meteorite impacts, giant volcanic eruptions, deep sea methane burps, sea level rises, sea level falls, rapid warming, rapid cooling, and so the list goes on. But, in my view, each of these speculations misses the point. That is because each speculation, on its own, usually fails to take account of the much bigger picture. For an appreciation of this bigger picture, we need to return to the library.

Cycles of calamity

Poring through the pages of books and papers on *Nummulites*, it is possible to build up a much fuller picture of their evolution through the Eocene. They had only started to thrive some 5 million years or so after the great extinction which took place at the end of the Cretaceous, 65 million years ago. This whole cycle—from calamity at the end of the Cretaceous, to giant protozoans in the Eocene, and thence back to calamity again—took a very long time by any reckoning: some 30 million years or so.[32] But there is something even more unsettling to report here.

Using the same rules for tracing the evolution of foram symbioses through deep time, we can discover that these great cycles—from calamity to calm and back to calamity again—have taken place numerous times over the last 300 million years. And one or more of these cycles can be traced within no less than forty separate lineages. Not only that but they have always followed the same tragic trajectory. They have always resulted in the total collapse of symbiotic forms, and by inference, of symbioses themselves. Ecosystems, it seems, will happily turn themselves outside-in, by taking in symbionts. But then they will rapidly turn themselves inside-out again, by losing them completely, followed by extinction of the host. It is an alarming pattern. Indeed, it is so regular, it is almost a prophecy.

It may be helpful to recall here the ideas of the social historian Elman Service who once wrote about complex civilizations.[33] Think, for example, about the Minoans, the Mayans, or the Romans. Such civilizations have depended upon human institutions—libraries, lawyers, scholars, and bankers. Elman Service argued that every big and complex human institution that has ever existed in the past has always fallen in the fullness of time, if we look at scales of hundreds or thousands of years. Now it could be argued that societies and ecosystems are not so very different. Each of these is a complex system having a hierarchical structure, complete with numerous nodes and connections. So here comes our big question. Could it be true that every big and complex ecosystem that has ever existed—including those of reef symbioses—has always collapsed as well?

The fossil record is arguably the only place to turn for answers here. And this seems to show us a terrifying pattern. Collapses in reef ecosystems have almost been the leitmotif of the later fossil record. In some cases, it took almost 5 million years before reef symbioses were able to emerge again after a major calamity. These reefal downturns are truly the dark ages of modern biosphere evolution. Geologist's names for these intervals of darkness ought therefore to bring terror to the tentacles of any aspiring polyp—ages called the Fammenian, the Scythian, and the Danian.

The scent of iridium

Soon after my arrival in Oxford in 1988, I bumped into a famous American geologist called Walter Alvarez. He was standing amid a gaunt assembly of dinosaur skeletons in the Oxford Natural History Museum. And he was explaining to a television cameraman about the ways in which he, and his father Luis, had transformed our thinking about the death of the dinosaurs, and about mass extinctions in general. Indeed, he had done exactly that, by means of a rogue element called iridium.

Walter Alvarez had been examining a section of rocks exposed along an alpine road at Gubbio in northern Italy. The cliffs along this pass exposed layers of seafloor that spanned the famous Cretaceous–Tertiary boundary. In the rocks below the boundary layer were fossils of Cretaceous plankton—mainly forams. Then, at a certain point in the road cutting, these all died out abruptly, to be followed by a thin boundary clay, and then by some completely new forms of foram plankton. These new forms could be seen evolving gradually from smaller to larger in size, and from simpler to more complex in architecture. No doubt about it. These changes could be followed around the world. There had been a complete revolution in the shapes and types of marine plankton. And it had taken place very abruptly, in just a few centimetres of ancient deep sea sediment.

Intrigued by these changes, Walter Alvarez sent samples to his father for analysis. Luis Alvarez was a famous chemist, and he started to look at the distribution patterns of rare Earth elements. One of these was iridium—a very rare element indeed here on Earth but rather more abundant in outer space. As is now rather well known, this iridium showed something rather unexpected—a rapid increase in its abundance coincident with the mass extinction of plankton. To the Alvarez pair, this could mean just one thing. A spectacular and irreversible collapse in the plankton had been brought about by some kind of extraterrestrial influence—perhaps a massive collision from a meteorite.

Scientists were understandably eager to put this dramatic hypothesis to the test. That is precisely what scientists do best. Sure enough, they found the fabled iridium spike at this same level, all around the world. They also recovered signals consistent with a huge meteorite, in the form of tiny globules of impact glass, or grains of quartz full of shock fractures, from within the boundary layer itself. Finally, of course, there came the evidence of a hidden meteorite impact crater, found on the Yucatan peninsula in Mexico at Chicxulub. It lay at just the right level, and it seemed to explain everything.[34]

The big ten

Fascinated by the extraterrestrial hypothesis of a bolide impact, the world of palaeontology began to explore other episodes of major biological turnover. Some have talked of a 'Big Five'. Others assume there are seven big events. But there are at least ten of them, stretching from the end of the Precambrian—with the extinction of the Ediacara biota—right up to the present day and the extinction of the monk seal and the woolly mammoth.[35] In fact, it was these characteristic changes in fossils that led Victorian geologists to distinguish the ten or so geological periods that we still use today. As students, we had been obliged to learn these periods and their subdivisions in much the same way that chemists had to learn the elements of the Periodic Table.

Interestingly it was a geologist who later became the first Professor of Geology at Oxford who first grappled with some of these major tuning points in the fossil record. John Phillips was the nephew of that famous canal engineer, and 'father of geology', William Smith—the maker of the first big geological map.[36] And it was John Phillips who first codified the fossil record into three main chapters: Palaeozoic, Mesozoic, and Cenozoic—meaning first, middle, and modern life. By the time that I was a student, the puzzles of the dramatic changes from Palaeozoic to Mesozoic, and again from Mesozoic to Cenozoic were justly famous. But the long, uniformitarian fingers of Charles Lyell still had the geological world in an icy grip. So it was still widely assumed that the biology of the planet had changed gradually not abruptly. There was to be evolution, but no revolution.

By the early 1980s, however, such gradualistic explanations were looking rather old and dusty. Walter Alvarez and his bolides were therefore well placed to set the world alight. Fossils ceased to be mere objects used by geologists for the dating of rocks—little more than magic beans. Fossils started to take on a life and a death of their own. Although the study of ancient ecology using fossils—called palaeoecology—had been in progress for a decade, it had been lacking a 'big question', so its achievements had seemed rather anodyne. But now a respectable question had at last come to hand. A very big question indeed: who were the winners and who were the losers in this geological poker game?

To answer such a question satisfactorily, the discipline needed to 'go global'. So there followed a race to discover the most complete and telling rock successions around the world. And a race to assemble a winning narrative about each and every downturn in the history of life. Such a wide field of enquiry also encouraged an explosion in expertise. Every kind of human combination began to emerge. Astronomers met up with starfish gazers. Bolide men discovered a latent love of botany. Cosmic ray physicists crammed up on their corals.[37]

In this way, gradualistic explanations were rapidly replaced by catastrophic ones. First of all, let us take a quick look at some of these strange happenings in the rock record, laid out as they may now appear to us.

This list of disasters needs to be read as though engraved onto a block of black granite, of the mournful kind to be met with in a Garden of Remembrance, or, indeed, in a War Memorial, which is exactly what this chapter is about.

The Ediacaran rocks, some 630 to 542 million years ago, contain the remains of the earliest large body fossils, such *as Charnia* and *Dickinsonia*.[38] Some have even suggested that their big rubbery bodies were those of giant forams, or that they cultivated symbiotic algae in their tissues.[39] Near to the end of this period, we find the earliest complex reef associations, provided with the ice-cream cones of *Cloudina* and the conker-like forms of *Namacalathus*.[40] The shocking story here is that all of these highly innovative forms disappeared near to the very end of the Ediacaran. This was arguably the first major extinction of animal-like organisms in the fossil record, and it involved the collapse of the earliest reef ecosystems.[41]

A radiation of new kinds of animal fossil soon followed, during the so-called Cambrian explosion. It was during this interval that all the major groups of animals first appeared, ranging from anchored sponges to burrowing clams, and from crawling trilobites to the first schools of fish. And no fossil group marks this explosion of animal body plans with more finesse than does a group of chalky sponges called the Archaeocyatha. These, along with calcareous algae, were the great reef builders of the early Cambrian, spreading globally into all manner of places with warm, clear, shallow waters. And it was here that they built lacy skeletons that are marvels of mathematical precision. Like modern reef sponges, it seems likely that these archaeocyaths cultivated cyanobacterial and algal symbionts within their tissues, although this has yet to be confirmed. But all their delightful designs came to nought. Some 520 million years ago, they were flourishing in the seas of the tropics. But 10 million years later they had disappeared completely. Their like would not be seen again for many tens of millions of years.

During the succeeding Ordovician and Silurian periods, reefs thrived again, but this time they were mainly built by now extinct kinds of cnidarian polyp. Some had a rough outer skeleton and are therefore called rugose corals, from the latin *rugae* for roughness. Others had lots of table-like internal platforms and hence are called tabulate corals. From their patterns of growth and symmetry, we can see they were only distantly related to modern reef-building kinds. These early coral reefs were soon joined by a novel form of calcareous sponge, called stromatoporoids. These were not delicate and cone-shaped like the Cambrian archaeocyaths but massive and lumpy, strewn about the seafloor of early reefs like cushions scattered across the floor of an Ottoman palace.

By the early Devonian, all seemed to be going well for these organisms. Single-celled forams were even starting to experiment with calcareous shells, and then with the cultivation of algal symbionts, as seen in forms such as *Semitextularia*. But then calamity struck. At the end of the Frasnian stage, these complex reef ecosystems collapsed completely. Corals were greatly reduced in number and diversity, while stromatoporoids and semitextulariids were destroyed for ever.

Reef ecosystems spent much of the following Carboniferous period in recovery. Little by little, though, new forms of symbiont-bearing protozoan—those fusulinid forams—began to flourish. The latter grew ever larger and more abundant towards the end of this period. Passing into the succeeding Permian, these fusulinid forams were evolving so rapidly that they provide us with clues for the passing of Permian time. These protozoan shell banks also began to stretch around the shores of the great Tethys ocean. Beside them in these Permian reefs there sat brachiopods—lamp shells. Modern examples are shaped a bit like a Roman oil lamp, and are tethered rather cautiously to the seafloor where they keep out of sight. But these Permian reef-dwelling examples—called richthofeniids—developed into baroque shapes, resembling horns of plenty. Such cornucopian shapes may have allowed them to expose their soft tissues to sunlight and to cultivate the algal symbionts within. Highly specialized ecosystems of this kind reached their peak in the middle of the Permian, but then started to decline in a series of steps, reaching a crescendo towards the end. The tally of death at the very end of the

Permian is shocking indeed. It is the largest known from the whole of the fossil record.[42]

The aftermath of this destruction was followed by a prolonged dark age, lasting through much of the Triassic period. But the space created by extinction allowed new forms of corals—the modern scleractinians—to evolve, as well as new forms of reef-dwelling foram called involutinids. The latter had seemingly achieved symbiotic shapes by the end of the Triassic. Needless to say, their development was cut short by yet another mass extinction event, at the very end of this period.

During the Jurassic and Cretaceous, we find reefs built not only from corals and sponges, but also from giant clams built like chimney pots. These reef-building 'rudist bivalves' may have cultivated symbionts within their mantle tissues, much like the modern giant clam *Tridacna*. These were joined by a succession of symbiont-cultivating forams. Major mortalities took place in these groups during the middle Cretaceous, and again at the end of this period. The end Cretaceous extinction event not only wiped out the dinosaurs, pterosaurs, and mososaurs, but also the ammonites, huge symbiont-cultivating molluscs,[43] most of the plankton, and some truly monstrous Foraminifera the size and shape of a Havana cigar, called *Loftusia*.

The succeeding Danian dark age was relatively short, just a few million years. And it was followed by yet another golden age, in the Eocene with its *Nummulites*. As we have seen, this was followed by collapse at the end of this interval. There followed a period of ups and downs in the Miocene. And from there we come to the current phase of reef ecosystem loss, with its missing monk seals and manatees. And coral bleachings.

Had all these ancient reef dwellers been warriors, we might have been tempted to write some final epitaphs. Words such as these come readily to mind:

> In Memoriam.
> Too young to die.
> Called before their time.
> Gone but not forgotten.
> They gave their lives so that we might be free.

That would all be humbug, of course. Our planet is without memory, unless we count the rock record. But we humans can still learn the lessons of all these genes gone missing in battle. Maybe.

The trigger

For many in the late 1980s, the nature of the narrative was starting to become clear. Walter Alvarez was right. Mass extinction at the end of the Cretaceous had been caused by a brush with a bolide, a huge meteorite some 10km across. Its impact had then caused dramatic warming, atmospheric toxicity, and mass mortality. Many groups of organisms never recovered. Including those living in the reef ecosystems.

All this was fine and dandy. Or it would have been, had there been no other signals at the end of the Cretaceous. For there were some conflicting signals too—such as massive volcanic eruptions in India over the same period (the Deccan Traps), as well as a sharp decline in animal diversity that—very inconveniently—began before the bolide arrived.[44] In fact, the more evidence we gained through the early 1990s, the more confusing the bigger picture seemed to become. At times, it felt as though there were as many explanations for all these troubling extinctions as fossils to measure them by. There were bolides and death stars; comets and supernovae; rifting continents and volcanoes; ice ages and sea levels; oxygen and methane; viruses and parasites, and so the list rolled on.[45] One could be forgiven for thinking at this time that just about everything beneath and above the sun has been invoked.

But it is here that we need to pause and reflect. When we reach a situation such as this—in which every extinction event has its own favourite trigger—and where no single kind of trigger can explain *all* the mass extinctions—then something may well be wrong with the approach itself.

The new approach, we need to remember, had involved looking for measurable physical triggers in the rocks. Often, they were sought out because geology departments had invested hugely in expensive machinery, and these machines were hungry for chemical samples. This approach also meant looking for triggers that were in direct proportion to the magnitude of the mortality. Hence 'big' causes were sought out for big extinctions, like bolides at the end of the Cretaceous. And 'little' causes were sought for lesser extinctions, like cooling at the end of the Eocene. But it gradually began to dawn on some of us that this approach was rather blinkered. Such a change of heart emerged from our growing understanding of the structure and behaviour of complex systems, including ecosystems—involving symbioses themselves.[46] And it emerged as well from a shocking realization that an old dictum of Charles Lyell—that 'the present is the key to the past'—might be flawed.[47]

We shall need to return to this puzzle again. But for now we shall press on with a related question that may help us gain a proper perspective. Could there have been a time, long ago, when the ways of the world were rather different? When those cells and their symbioses did not collapse so easily? To answer this, we must now take a trip into the outer darkness of Deep Time.

CHAPTER 9

GUNFLINT

Schreiber bound

As we raced between islands and then sliced across open water, the wake of our little launch, called *l'Oiseau Bleu*, engraved an almost perfect arc onto the smooth blue liquid of the lake. Even Lake Superior, we learned, can have its calmer days. The sky above us seemed sublime, and the crew of the launch seemed ready for action, too. That was easy to achieve, of course, with a crew of only one—the pilot. He was a French-Canadian boatman with a big straw hat and a kind offer to take us wherever we wanted. And on that we were clear. We needed to gain a better context for the symbiotic origins of complex eukaryote cells. We yearned to see for ourselves the place where the search for early life on Earth had really begun. And, then again, to learn about two men who had started this quest—Elso Barghoorn and Stanley Tyler.

Clutching at the boat rails, my students Leila and Sean started to lean forward, scanning the horizon for any signs of a beach.[1] The shoreline around Lake Superior, near to Rossport in Canada, is liberally sprinkled with spruce-clad islands, yet we were looking for a very special beach. It was marked by an isolated islet that stood like a sentinel, away from the shore. We were looking out for some gentle rocky slopes as well, backed by a dark canopy of pines (Plate 11). After a while, the launch span round in its final arc, the roar of its engines dimmed and we could hear little more than the lilting of the lake against the hull. To our left, indeed, was an offshore island that fitted the description. Glances were exchanged. Nods were given. We had finally reached the locality. The famous Schreiber beach locality.[2]

From a distance, the shoreline appeared to be covered in black rocky lumps, each about the size and shape of a large pillow. They were not exactly like bedroom pillows, though, because these were big, black, and slimy. Such slimy lumps were the remains of a very ancient lava flow— submarine andesite and basalt that had long ago hissed and splintered onto the seafloor. Plump pillows just like these can be seen forming in places like Iceland today, where iron-rich lava erupts.

Once ashore, we could see these pillows of black lava piled up along the shoreline.[3] In places, they had gained steely grey coatings, resembling the layers of glazing on a cake. Bending down more closely, we could see that these layers were wrinkly, with some being as pale as glass, while others were as dark as iron (Plate 12a). It was within these pale glassy layers that something marvellous had happened long ago.

As we stepped ashore, I tried to imagine the scene—a coastline of ancient pillow lavas and pebble beds, already a billion years old. The shore was being splashed by waves. But this ancient sea—1900 million years ago—was not rich in chalky matter, like those of the tropics today.

Instead, it was surprisingly rich in glass.[4] Much of this glass—more properly called chert or cryptocrystalline silica—had come from the weathering of granites and other continental rocks. In solution, it had then been transported down rivers to arrive in the sea. Here it reached concentrations far above those found in seas today. The stage was now set for a crystal garden to flourish.

Over a short period of time, microbes on this ancient seafloor had started to grow, then to grow up, and then to have progeny. Soon, they began to bloom prolifically, especially around the pillows and boulders of the shoreline. It was here that glassy material—perhaps assisted by flows from hot springs—began to settle out of solution and to form a gelatinous ooze that tightly hugged the pillows. As it congealed, it also started to entomb some unsuspecting microbes as well. Not aggressively like the ash of Pompeii, but very gently indeed. Some of the microbes had been resting. Others were squirming or dividing. Many were washed in by waves. But cell-by-cell, their brief lives came to an end. For us, though, their story had only just begun.

The quest

Schreiber Beach is a lonely spot that seems well fitted for studies of early life and the deep history of the cell. How marvellous it must have been to make such a discovery and to go down in the history of human thought—at least as a footnote.[5] This is something that nearly every palaeontologist dreams of—making a great discovery about the early history of life, one that completely changes our picture of 'how things began'; or one that converts its authors into legend. Needless to say, I was eager to find out the rules of the game.

It is rather unusual for day-by-day evidence about such a great discovery to survive. We hear a bit about this with Charles Darwin and his fabled Galapagos finches, of course.[6] We also sing the praises of Howard Carter, with his talk about 'wonderful things' while opening up

Tutankhamun's tomb.[7] But what about great fossil discoveries like those of the famous Gunflint chert biota—how on Earth did that ever come about?

One day in Amherst in Massachussetts, some of the evidence for this came to hand. Lynn Margulis kindly put in front of me a little known collection of letters. They were letters that dated back to 1951. Some of them were written by Stanley Tyler to Elso Barghoorn, and others were written in return by Elso Barghoorn to Stanley Tyler. Both of them are now highly respected—and world famous—in the annals of fossil hunting. At the time of the letter writing, Tyler was a Wisconsin field geologist, with a special interest in coal and iron deposits and Barghoorn was a professor of palaeobotany—the study of fossil plants—with a special interest in coal fossils. He had gained a position at Harvard University shortly beforehand.[8] What might their old letters tell us about the rules for making a great scientific discovery? Could there be a protocol to follow here?

As I poured through the correspondence, a ghostly pattern began to emerge. If you happen to have a pencil and paper to hand, you might wish to write these rules down for yourself. For it emerged that these are the golden rules of the game:

1. Look in the wrong place.
2. Look at the wrong things.
3. Look at the wrong time.

I was delighted by this evidence, of course. For these are the very rules I have been following—to the very letter—all my life. Let us take a closer look together at how these three strange rules guided both Barghoorn and Tyler towards the Hall of Fame.[9] But first of all, we need to return to the shores of Lake Superior and take a closer look.

Peering more closely at the rocks, we could see again those cherty layers draped around the lava boulders. But careful inspection showed that they were very far from uniform. Instead, they were arranged in bands of white, grey, and black, seldom more than a few centimetres thick. In places, an even stranger patterns could be seen in the rocks—lots of little domes the size of thimbles, arranged side by side and stacked on top of one another. In section, this gave to the rock a rather cabbage-like appearance. On the seafloor, they must have looked like a huge dish of cauliflower cheese.

These strange structures are called stromatolites. They have for long puzzled geologists. Such things were first reported and illustrated as early as 1825 by a long forgotten geologist called J. H. Steele.[10] Splash zone stromatolites were seen by Charles Darwin when his ship *HMS Beagle* landed on St Paul's island in the middle of the Atlantic in 1832.[11] And stromatolites cropped up again when James Hall studied the rocks of New York state in 1883.[12] By this time their cabbage-like shapes were thought to have been constructed by forams or sponges. But in 1908, a German palaeontologist called V. Kalkowsky advanced a much more controversial hypothesis: that the domes of stromatolites were actually constructed by algal cells on the seafloor. At first, this idea was met with scepticism. But doubts about this began to dwindle after 1933. It was in that year that a geologist from Cambridge in the UK came up with some evidence. Maurice Black had been paddling across the tidal lagoons on the island of Andros in the Bahamas[13] when he noticed that dense growths of cyanobacteria (then called 'Blue Green algae') were forming an 'algal mat' that seemed to survive by growing upwards through the layers of sediment. This, and the stickiness of their cells, ultimately led to the trapping and binding of cabbage-like layers of sediment. Stromatolites were actually forming today. And they had a microbial connection.

So, by 1954, the biological potential of stromatolites had been suspected for decades. Indeed, they were known to abound within rocks of similar age across the lake in Minnesota.[14] A search for the cells of microbes within stromatolites of the Gunflint chert—and its equivalents in Michigan and Minnesota—would therefore have seemed like a very logical way to begin.

So did Barghoorn and Tyler begin by looking inside stromatolites from the Gunflint chert at Schreiber beach? Not a bit of it. The word 'stromatolite' was never mentioned in their paper of 1954. Those authors began with coal.[15]

The Michigamme slate

At the time when the biota in the Gunflint chert was discovered, Elso Barghoorn had been cutting his teeth as a specialist on fossil plant remains, like those being exhumed from American coal measures in Pennsylvania. That was a valuable skill in the industrial age. The running of steam engines and battle ships, indeed the winning of wars, needed vast and dependable reserves of coal. So coal was to become a strategic resource to the United States as it had been earlier to Britain. Lots of coal required lots of coal-geologists.[16] Stanley Tyler, on the other hand, was not really a coal man at all, nor a fossil man like Barghoorn. He was a field geologist from Wisconsin, working mainly on iron-rich rocks from around the Great Lakes.

The Gunflint story really begins with a letter on 15 February 1951. Tyler sent over a brief message to Barghoorn in the Botany Department at Harvard—giving details of a specimen of rather strange coal that he had helped to discover in Michigan, from the Iron River region.[17] He was writing to Barghoorn in Harvard because there was a curious enigma about this coal. He had not expected to find a layer of coal in rocks much older than 400 million years or so—rocks of Devonian age. It was only from the Devonian and Carboniferous periods that land plants had started to appear in the fossil record. Those early lands plants produced marshes and forests, and ultimately coal swamps. In a world before coal swamps, there should have been no coal.

But this old coal from Michigan was many times older than the coal measures of Pennsylvania. It came from a black shaly deposit within a much deeper geological unit, that Tyler called the Michigamme Slate.[18] The presence of coal in rocks so ancient was a big puzzle because life had never been confirmed in rocks as ancient as those from Michigan. There were hints of life, of course, in the form of highly squeezed and cooked veins of ancient carbon known as graphite,[19] and also in the form of the

Figure 26. Elso Barghoorn (1915–1984) taking shelter from rain during work on the hot springs of New Zealand. This image from the Waimangu thermal area on the North Island New Zealand was taken by Cecilia Lenk in June 1980, and was kindly provided by Paul Strother.

stromatolites we have just met. But nothing definitive had ever been confirmed. That is why Stanley Tyler dashed off his letter, in the hope of hearing back from the Harvard professor (Figure 26).

Elso Barghoorn was clearly intrigued to hear of this report coming in from the wilds of Michigan. But in the months that followed, there was a sense of puzzlement in his mind as well. Could the Michigamme coal really be so ancient? And was it really coal at all? Attempts were therefore made to discover whether this coal-like rock had really come from the Michigamme Slate, as Tyler had thought. Could it be that the sample had been accidentally transported there from the coal measures of

Pennsylvania, for instance, perhaps by glaciers during the ice age? Or, worse still, by mineral trucks during the industrial age? Or worst of all, could the sample have been carelessly mislabelled? Might their beautiful story be destroyed by an ugly little fact like that?[20] Happily, by October 1951, such fears were starting to melt away. Analyses in the laboratory were beginning to show that Stanley Tyler's mysterious sample was not only more baked and crushed than coal from the Carboniferous, but it had a different chemical fingerprint too. These coals were very ancient. Now that was curious.

The Hooke connection

In looking for cells within coal measures of the Carboniferous, Barghoorn was following firmly in the footsteps of Robert Hooke. As we have seen earlier, the first discoveries of cells in fossil coal were made by Hooke, back in 1665. Both charcoal and coal were found by him to be constructed from cells. Indeed, Hooke provided the first ever reports of fossilized cells preserved in rocks, writing about them with almost breathless excitement.[21] Most intriguingly, Hooke even conducted the earliest experiments on fossilization, including the petrifaction of cells in wood.[22] By the early nineteenth century, it was thought that cells within coal plant fossils were like those found inside living ferns and pine trees. That was a not an unreasonable conclusion before the facts of evolution were known. Little by little, however, this argument for homology began to collapse. Microscopists like Robert Brown were starting to find that the cells of coal plants were nothing like those of modern ferns and pine trees at all. They belonged to wholly extinct groups of plants.[23] How strange all this must have seemed in the world before Darwin. What reason could any creator have found for extinguishing all those harmless arboretums?

By 1951, ideas had changed markedly. But we need to remember that the living world was still divided by biologists into two big kingdoms. A living thing was either a plant and stayed put, or it was animal and moved about. It is true that the outlines of plant evolution had become quite well established by that time, mainly along Darwinian lines. Indeed, the story of plant evolution provided some of the most architecturally elegant examples of evolution in the fossil record.[24] However, rather little was understood about their evolution, and that of fungi and microbes, in the world before the Devonian Period, some 400 million years ago.

Perhaps because of these uncertainties, we find no great sense of urgency in the initial series of letters that passed between Barghoorn and Tyler during 1951. Barghoorn was no bombast. Nor would a bombast have gone down well in botany. Almost a year was to pass, therefore, before his suite of rock slices became available, of a kind suitable for examination down a microscope. As soon as those slices appeared, Elso Barghoorn got moving. He began to have his samples exposed to X-rays, or dunked into a succession of very strong acids—nitric, hydrochloric, and even boiling hydrofluoric. Finally, he put them through their paces using various optical tricks to get at their deep underlying structure and composition. At long last, exciting patterns began to emerge.

While lighting his sample from above, Barghoorn thought he detected tiny fan-like markings. They reminded him of tiny tussocks of a Blue Green alga called *Rivularia*. This discovery seems to have stirred his imagination. From May 1953 onwards, the pair began to think about writing up their results. 'In writing my section of the paper,' wrote Barghoorn 'I should like to point out that our Michigamme shale deposits are among the oldest organic sediments representing definite evidence of algal life which have yet been found [in the history of life on Earth], and for which the evidence is not too controversial.'[25]

So did Barghoorn and Tyler write up those coals from Michigan, first hinted at in Tyler's letter of 1951? Not a bit of it. These coals were never mentioned in their premier work.[26] Those two authors had found something altogether more exciting. And that something takes us over to Scotland.

Through the looking glass

One day in 1912, a Scottish country doctor called William Mackie spotted some strange looking cherty rocks built into a stone wall, in a field at Rhynie near Elgin, on the southeastern edge of the Scottish Highlands. Curious, he sliced those rocks so thin that they could be examined under a microscope. What he saw was later to astonish the world. The remains of stems and roots from the oldest known land plants had been preserved exactly where they lived, still standing upright along the margins of a now extinct thermal spring.[27] Better still, their plants were preserved at the cellular level. It was very much like looking through a lens at a modern botanical preparation—except that these plants and animals were some 400 million years old.

This famous deposit—known as the Rhynie chert—was lovingly described and illustrated from 1917 onwards by Robert Kidston and Henry Lang.[28] Beneath their microscopes they pointed out roots, stems, and spore-producing bodies, often complete with their cell walls and cell contents. Amazingly, some small animals were later found entombed within the chert as well. These included some of the oldest known fossil insects, caught eating away at the earliest spores on the earliest land plants.[29]

It was the charm of the Rhynie chert that gave to Stanley Tyler his grand idea to start looking at chert within the banded iron formations themselves. Could the cell structures have stayed pristine in the cherts, like those from Rhynie? Had they perhaps been embalmed within the silica oozes very quickly indeed, and thereby protected from the crushing overburden of three miles of sediment? Tyler decided to take a closer look.

The lengthy business of making thin sections came first. The rocks had to be collected, curated, sliced with diamond-tipped saws, polished with corundum paste to exactly thirty thousandths of a millimetre thick, pasted on to glass slides, covered with cover slips, labelled, and then stored in trays. Such a process has seldom been undertaken without considerable cost. But, in his case, the rocks were arguably worth their weight

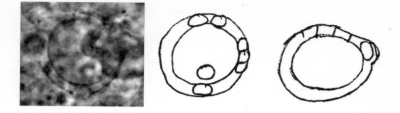

Figure 27. The Dawn Sphere of Stanley Tyler, later called *Eosphaera tyleri* by Elso Barghoorn. At left is shown an image down the microscope, showing a typical cross-section through the middle of a double-walled sphere, itself made up from tiny spheres. At right are shown two of Stanley Tyler's rough sketches, taken from his letters to Elso Barghoorn. Fossils are about 20 microns wide.

Note. These images were kindly provided by Lynn Margulis and Dave Wacey.

in gold. Ovoid cells started to appear in abundance within the Tyler Gunflint cherts, at first from 'the Iron River locality'. These structures were later called *Huroniospora*.[30] Next, Tyler came up with colonies of tiny cells arranged in a sphere around a hollow centre. These were later to be called *Eosphaera* (Figure 27).[31] Yet another resembled a tiny black star-burst—later called *Eoastrion*.[32]

During this initial phase, Tyler got to grips with his cherts but Barghoorn stayed loyal to his coals. By June of that year, Tyler had even begun searching along the northern shores of Lake Superior, some 30 miles east of Port Arthur in Ontario. And by October, his most important deposit had been discovered along the shoreline near to Schreiber, further to the east, which is exactly where this chapter began.[33] Tyler's new material was incredibly rich—it contained up to 1.5 million micro-fossils per cubic centimetre (Plate 12b).[34] Writing excitedly to Elso Barghoorn, he noted: 'This section contains filaments which range from under one to over two microns in width. The filaments in general seem to be non-branching and in some cases appear to be connected to ovoid bodies. This material is the best I have ever seen—perhaps you can do something with it.' There followed a pencil sketch with the handwritten note 'I believe this material is probably blue-green algae'.

From here onwards, there was an evident thrill about the fossils they were now finding. Filaments began to receive fine compliments. One kind reminded him of a living relative of the Foraminifera, called the Radiolaria. This little starburst was later called *Eoastrion*. (Figure 28) Another strange form was said to resemble a tiny jellyfish, later called *Kakabekia*.[35] 'This [slide] is a lulu,' cooed Tyler, barely able to contain his excitement. His little 'jellyfish' was not an animal fossil of course, it was a kind of fossil bacterium. But it was the most complex looking thing that had ever been seen in such ancient rocks. That is arguably true, even today. And its discovery marked a change in the working relationship between Tyler and Barghoorn too, because they moved swiftly from formal manners to first name terms.

By November, Elso was starting to clamber out of his shell. 'The new material you have sent me is best described as "sensational". I think we have something that is certainly unique in recent discoveries in paleontology, and I only wish I could drop everything else and go to work full time on our pre-Cambrian flora.' It was rather like finding a new continent, said their fossil-hunting friend, Robert Shrock, a shade enviously.[36]

Apart from Shrock, these beautiful fossils were being kept secret at this stage. This had some advantages—their ideas could not be stolen or

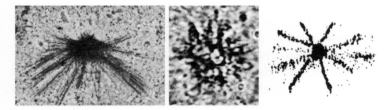

Figure 28. The Dawn Star of Stanley Tyler, later called *Eoastrion* by Elso Barghoorn. At left are shown two images down the microscope, showing a typical cross-section through the middle of a radiating mass of bacterial filaments. At right is shown one of Stanley Tyler's rough sketches, taken from a letter to Elso Barghoorn. Fossils are about 30 microns wide.

Note. These images were kindly provided by Lynn Margulis and Dave Wacey.

copied. But it had some disadvantages, too. It made finding out things about the Gunflint chert a good deal more difficult. One worry, for example, was the actual age of this chert assemblage. Establishing an accurate geological date for a deposit—in millions or even billions of years—is always a crucial step for any study of early life. At the start of their research, the Gunflint chert was thought to date from around 1000 million years ago—nearly three times the age of the Rhynie chert and twice the age of the Cambrian explosion of animal life. But by the end of 1953, some new dates—making use of radiometric clocks—were coming to hand, and the age of their assemblage was about to double, from 1000 to almost 2000 million years.[37] In so doing, it had quadrupled the span of life on Earth that was known at that time. Such an ancient date feels amazing even now. It must have felt truly vertiginous back then. Barghoorn and Tyler clearly had a big and important story on their hands.

So did they get that story right in the end? Like that old story about the curate's egg, we can only answer politely that it was *good in parts*. As is often the case, it is those parts which the authors got wrong that shows us how science really works. To appreciate this more fully, we need to take a closer look at those addled parts.

A fossil trap

Especially difficult for these early pioneers of cellular life in Deep Time was the question of how to interpret their newly found fossils. As we have seen, comparisons with Blue Green algae—now called cyanobacteria—were being made by both Tyler and Barghoorn from the earliest stages of their research. And these allusions remained in their paper too. Thus, some of their filaments were compared with that wriggly Blue Green cyanobacterium we have met with, called *Oscillatoria*. That still seems a reasonable suggestion today. Other filaments were compared by them with a tufted cyanobacterium called *Rivularia*.[38]

Browsing through their correspondence we can also pick out some of the very curious speculations that were running through their heads back in 1954. As we have seen, umbrella-like *Kakabekia* was rather jokingly compared with a jellyfish. That suggestion was never intended seriously, because the fossil is far too tiny. *Kakabekia* was probably some kind of bacterium, being little more than a hundredth of a millimetre or so across. Comparable living examples have even been illustrated from the medieval sewers of Harlech Castle in Wales.[39] But we also meet with the names of organisms that nobody would expect to claim—and then get away with—in rocks of this age today, including a group of protozoans called the Radiolaria (Figure 29).[40] And two of these names in particular hold special interest for us here. For both are now known to be complex eukaryote cells.

The first really unexpected name that we meet here is that of a 'coccolith'. This name means berry-stone in Latin. These berry-stones are actually elegant little discs of chalk that today are secreted by a form of eukaryote cells with Golden Brown chloroplasts, collectively known as haptophytes. They are close relatives of the diatoms we met with on the Pedro Bank. Like diatoms, the chloroplasts of haptophytes have triple walls consistent with an origin from a Red algal symbiont—from a secondary symbiosis.[41] In the Atlantic, their plankton blooms form the hors d'oeuvres for fish such as herring and halibut. Fishermen even track these coccolith blooms with satellites. After these blooms have collapsed, the cells die and their chalky skeletons fall down to the seafloor, to form a blanket of chalky ooze. The Victorian biologist Thomas Huxley once marvelled at their proliferation in the modern oceans, and at the coccolith shells brought up with the dredges of *Globigerina* ooze by ships like SS *Great Eastern* during the laying of trans-Atlantic cables after 1856.[42] Huxley had even lectured at length 'On a Piece of Chalk' to a public still ignorant about the abundance of coccoliths within the chalk of England,[43] now known to be some 100 million years old.

Barghoorn was greatly intrigued by a resemblance he could see between some of his fossils and the forms of chalky coccoliths.[44] He even thought he had found something like their calcareous scales.[45] But, in the fullness of time, none of this has turned out to be true. Coccoliths have

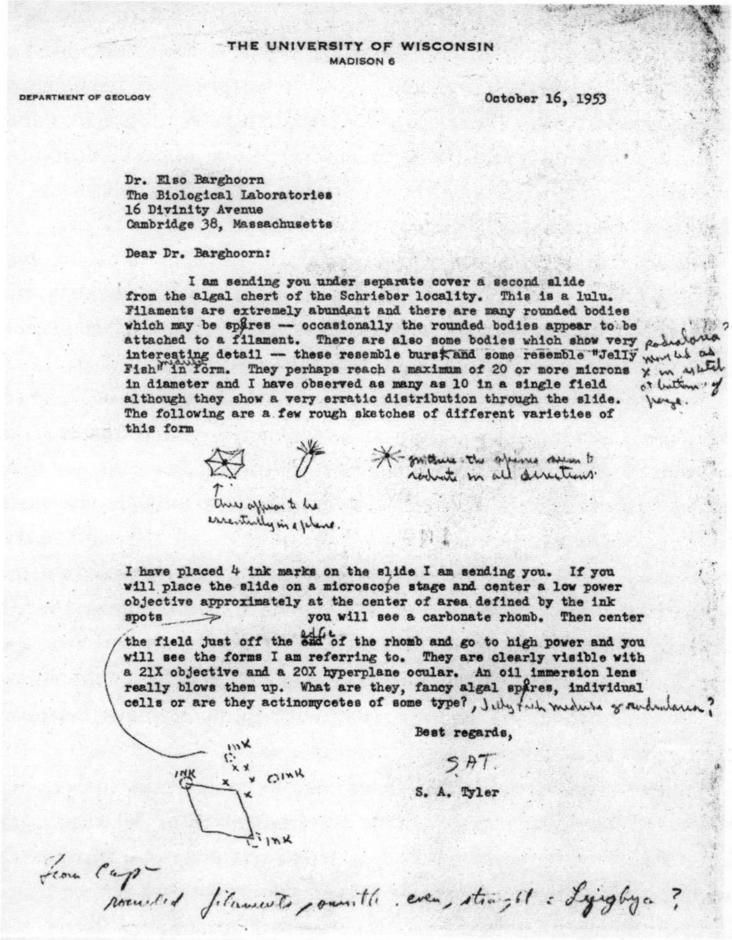

THE UNIVERSITY OF WISCONSIN
MADISON 6

DEPARTMENT OF GEOLOGY

October 16, 1953

Dr. Elso Barghoorn
The Biological Laboratories
16 Divinity Avenue
Cambridge 38, Massachusetts

Dear Dr. Barghoorn:

I am sending you under separate cover a second slide from the algal chert of the Schrieber locality. This is a lulu. Filaments are extremely abundant and there are many rounded bodies which may be spores — occasionally the rounded bodies appear to be attached to a filament. There are also some bodies which show very interesting detail — these resemble burs and some resemble "Jelly Fish" in form. They perhaps reach a maximum of 20 or more microns in diameter and I have observed as many as 10 in a single field although they show a very erratic distribution through the slide. The following are a few rough sketches of different varieties of this form

I have placed 4 ink marks on the slide I am sending you. If you will place the slide on a microscope stage and center a low power objective approximately at the center of area defined by the ink spots you will see a carbonate rhomb. Then center the field just off the end of the rhomb and go to high power and you will see the forms I am referring to. They are clearly visible with a 21X objective and a 20X hyperplane ocular. An oil immersion lens really blows them up. What are they, fancy algal spores, individual cells or are they actinomycetes of some type?

Best regards,

SAT

S. A. Tyler

Figure 29. 'This slide is a lulu'. This is the message penned by Stanley Tyler to Elso Barghoorn, in October 1953. It shows his reconstructions of fossils later called *Kakabekia* and *Eoastrion*. At the bottom, he ponders whether they could be jellyfish or radiolaria.

Note. This photocopy was kindly provided from the collections of Lynn Margulis. The originals are housed in the Barghoorn collections at Harvard.

been studied by scientists in their billions since 1954. Not one has been found in rocks older than the Triassic Period, before about 250 million years ago. The presence of Golden Brown haptophytes in the Gunflint chert—nearly 2000 million years old—would have been an astounding discovery. It would imply that secondary symbioses of algal chloroplast had already taken place by then. Alas for Barghoorn, the fossils he described seem to have been mere mineral growths.[46]

Equally surprising was Barghoorn's conclusion that one of his fila-ments was 'most certainly a fungus'.[47] We need to recall here that the fungi comprise a highly diverse group of complex eukaryote cells, famil-iar to us from toadstools and lichens, mushrooms and moulds. Having at most a single flagellum they are regarded as Unikonts like ourselves.[48] They can play a hugely important role in the modern world as well, not least because their spores and their filaments—called mycelia—tend to get everywhere. In no time at all, these mycelia push out digestive juices to break down organic materials—into jam or milk, cakes and cadav-ers—to adsorb their products. Barghoorn himself had even explored the fungal destruction of things like food, tents, and binoculars during World War II.[49] Fungi are also involved in a kind of warfare themselves, churn-ing out the antiobiotics we so cherish today, such as penicillin. Forget about Athlete's Foot and nappy rash. In truth, the living world thrives on the beneficent bounty of fungal infection.

The position of fungi in the Tree of Life is of considerable importance for our reading of the history of life in Deep Time. Modern work using DNA and RNA suggests that fungi lie close to the base of that branch within the tree that evolved into sponges, jellyfish, real fish, and our-selves.[50] Hence the earliest fossil fungus would be a finding of first class importance.[51] But not a single claimant for that title looks at all convinc-ing to us today—that is, until close to the end of the Precambrian, about 700 million years ago. Before that time, the world seems to have been without moulds or mushrooms. What Barghoorn had mistakenly regarded as 'almost certain' fungal mycelia were almost certainly just the slimy sheaths of bacteria instead.[52]

Not the least of their problems was the distinction between objects that were real organisms and those things that were simply mineral growths. Perhaps the most unfortunate mistake of their whole paper back in 1954 turns up in two of their four figures—showing clusters of spheres like bunches of grapes. At first, these were regarded as protective coatings around the algal cells. But we now know they were false signals—nothing more than mineral growths. While their mistake was naïve, it was wholly admissible because their science was still young. Few scientists would dare to make such a mistake today and hope to survive with reputations intact.[53]

Most surprising of all, though, is a name that we don't find mentioned at all here—that is, until near the very end of their letters—the name of *bacteria*. Nor is this word ever mentioned in their paper. That omission may seem bizarre to us now because 'bacteria' is what we would choose to call every one of these incredibly ancient fossils. But its absence was not bizarre at the time for one good reason—the unique place we now give to bacteria went largely unacknowledged until the work of later scientists such as Lynn Margulis and Carl Woese.[54] We also realize now that life does not have to be dropped into either the 'plant' or the 'animal' bucket. It can be bacterial, too. Or it can bring many of these things together inside a single parent cell—like the world of the corals, a world turned outside-in.

A mystery in microbes

Our perspective about life on Earth was destined to change after the publication of the paper by Stan Tyler and Elso Barghoorn in 1954.[55] The span of life was to treble, reaching half way back to the origin of the solar system itself. But while the Gunflint chert microbiota created a moderate stir when it was first published in *Science*, we can now see that it was not the first ever report of genuine cells from deep in the Precambrian. Surprisingly, the first report of superbly preserved cell clusters came in 1899, from phosphates in the Torridonian rocks of northwest Scotland, about a billion years old.[56] Nor was it even the first report of possible cells from the Gunflint rocks. That report appeared in 1922.[57] Nor was the paper of 1954 substantially correct in its conclusions, as we have just seen. These authors were just the first to report them with authority.

By the time of publication in 1954, it was Stan Tyler who had done all the running around, the collecting, the preparation of thin sections, and much of the referencing. He even spent a week around the hot springs in Yellowstone Park in Wyoming, trying to understand how silica deposits form.[58] Elso Barghoorn was at first caught by the inducement of a 'coal' deposit in the Michigamme Slate. Only after Tyler's discovery of the Schreiber Beach chert was he drawn in really deeply.[59] But Barghoorn never penned a full report together with Tyler at this time. All this is sadly moving.

Poor old Stanley Tyler. He loved his family and he loved their garden. He loved to ramble and he loved to fish. Most of all he loved his little fossils. For ten years, he dashed off excited letters to Barghoorn, full of little sketches. Or posted off samples in quick succession, full of ancient microbes. He wrote and he wrote. But little ever came of it during his lifetime. In October 1963, he took ill and died, worn out at the not very great age of 57. He was never able to bathe in the limelight his fossils would one day bring.

One warm and sticky summer's evening in a restaurant, set beside the London Natural History Museum in Kensington, I had the good luck to share an Indian curry with Andy Knoll—Barghoorn's heir and a Harvard professor too. I asked Andy to help me pull away the mask from Barghoorn the man. Why was Barghoorn seemingly so reticent? And why did he never produce a major tome together with Stanley Tyler on his findings?

On the up side, said Andy, the lectures that Elso Barghoorn gave to his students covered a vast canvas—ranging from recent economic history to the very origins of life itself. Many good students were drawn in.[60] He was a polymath and a catalyst, too. Quite a few of his students even went on to work on the Gunflint chert.[61] His research was clearly riveting and his images could effect a fascination. But there was a downside. First came the fact that 'Elso was not a people man'. He was shy and reticent, and hence chose to withdraw from any form of social gathering. Second, his voice was dry and he would deliver his lectures in a factual monotone that would tend to send some of his students to sleep.[62] And then I learned about some strains in his life, too—the result of his trying to juggle several personal tragedies. Not only his father but his second son and his third wife were all to commit suicide.[63] There was clearly some darkness in the mixture here. Last but not least, Elso Barghoorn was easily distracted.[64]

All this helped somewhat to explain why the original paper of 1954 was followed by little more than dribs and drabs over the ensuing decade.[65] When Stanley Tyler died in October 1963, Barghoorn was seemingly on the verge of letting the Gunflint fossils slip from his grasp. By this time, we need to remember that America's Sputnik Moment had already arrived. That cathartic moment took place in 1957. And this little bleeping Soviet satellite caused great consternation in America. By the middle of 1961, President Jack Kennedy had announced that his nation was going to the Moon. America would get there first, and then move on to Mars. All this required tough thinking about other worlds, and about the likelihoods of life on other worlds. By 1964, NASA was therefore on its way into space. The Moon had been surveyed from space. And the Gunflint chert and its biota was about to become a truly hot topic. What followed, though, was to come very close to tragedy. A professor from California called Preston Cloud was bursting to get into the limelight here. He poked and he prodded, he mapped and mused on the Gunflint fossils. Unable to contain his patience any longer, he submitted a very long, highly detailed, and extremely thoughtful paper to the journal *Science*.[66] Cloud, it seems, had even mapped the sacred Schreiber rocks for himself, after being marooned on the beach by a fierce storm.[67] And that star pupil called *Eoastrion* was made, by wizardry, to shape-shift into an iron-oxidizing microbe, resembling *Metalogenium* that lives today in freshwater lakes.[68] But an academic storm was also about to brew around him.

Barghoorn learned about this challenge through the editor of *Science* when he was asked to review the Cloud paper. Understandably, this placed him under huge pressure. All his hard work with Stan Tyler could have been rendered irrelevant. He therefore played for time by using a well known gambit. It is called 'sitting on a paper' after it has been received for comment. The rules are perfectly simple: (1) File the offending paper in a drawer; (2) write your own paper as quickly as you can; (3) submit your own paper; (4) now pull the original item out from the drawer, and write the review you were asked for; (5) whistle innocently, and hope your paper comes out first or, at worst, they both come out together. There are many fine precedents for this in science. It is precisely what Roderick Murchison had done to the young upstart Louis Agassiz in 1840 to forestall the latter's glacial theory. And it is what both Darwin and Lyell did to that young upstart Alfred Russel Wallace in 1858 to forestall his evolutionary theory. So this is what Bill Schopf tells us he did with Elso Barghoorn back in 1965. Those two *sat on* the Cloud paper for one month in order to plough through the huge compendium of Tyler's photographs. And then they sat on it for another month in order to allow Barghoorn to gather his thoughts together. Mercifully for our story, he succeeded in doing just that (Figure 30).[69] For such is the way of the world.

Why does all this matter so much to our story of symbioses in deep time, and the origins of the eukaryote cell? It matters greatly because the Gunflint chert, at nearly 1900 million years old, provides us with a major benchmark for our narrative. Despite the early thoughts of Tyler and Barghoorn, there is no convincing evidence for the presence of complex eukaryote cells like those we have met with in preceding chapters. Earlier suggestions that the absence of eukaryotes from the Gunflint chert is a result of strange environmental conditions will no longer wash because they are absent from deeper water cherts as well.[70] Recent claims for chemical 'sniffs' of eukaryotes—molecules called biomarkers—in much more ancient rocks, some 2700 million years old,[71] have since been withdrawn because they appear to have been much younger contami-

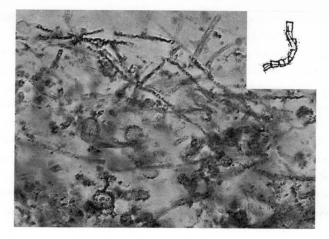

Figure 30. Filamentous and rounded microfossils from the Schreiber beach locality of the Gunflint chert. The filaments include forms assigned to *Gunflintia* by Barghoorn. The rounded, or coccoid, forms were assigned to *Huroniospora*. The inset shows the rough sketch of *Gunflintia* made by Stanley Tyler in his letter to Elso Barghoorn. Field of view about 100 microns wide.

Note. I am grateful to Tom Hearing and Lynn Margulis for these images.

nants.[72] Large spherical microfossils from rocks as old as 3200 million years lack eukaryotic features and are probably organic-walled envelopes of bacteria.[73]

No Reds, Greens, or Browns, and no coccoliths or dinoflagellates have yet been found in really ancient rocks.[74] If we are going to find the origins of complex cells, it seems we may have to look in rocks a bit younger than 1900 million years.

All this matters greatly, too, because it also tells us about the rules for making a great scientific discovery. It is time to return to those rules from the start of this chapter:

1. Look in the wrong place.
2. Look at the wrong things.
3. Look at the wrong time.

Examine these words very carefully because, like any good riddle, the paradox contains its own answer.

Is it possible that we have been looking at this list in the wrong way? Indeed it is. We can learn a lot about life when we realize we have been asking the wrong questions, or by looking at things in the wrong way. Tyler and Barghoorn began by looking in the wrong place, in Michigan. They began by looking at the wrong things—in the Michigamme coal. And they were looking at the wrong time—before the biology of bacteria was properly understood.

But that is not quite my point. The answer to our riddle is spelt out clearly in the list—maybe you haven't seen it yet. The clue lies in the word 'Look'. Repeated three times.

CHAPTER 10

CARPENTARIA

The 'Boring Billion'

'Oh Goodness! This takes me right back to the Moon!' said a large bear of a man standing beside me. It was a strange thing to overhear deep in the Australian outback. But this was no ordinary geologist. I was working in the field with vintage NASA geologist John Lindsay. By 1998, almost three decades had passed since my voyage around the Caribbean. I had heard that John had helped to train up lunar astronauts, including John Young and Jack Schmidt, for the Apollo missions back in the early 1970s. He was a pioneer in geological sampling and the mapping of the Moon. He had even written his own standard textbook on the sediments of the Moon.[1] To be honest, he would never stop talking about the Moon and his Apollo missions. John, it has to be admitted, was Moon-struck.

The thing I valued most about John Lindsay, though, were not his lunar qualities but his Australian ones. Cheerful by nature, and with twinkling eyes set above a large and jovial frame, John had grown up as a farm boy, conditioned to despise all forms of humbug. And he especially despised humbug in high places. He was tireless in his pursuit of truth—even if it was sometimes his own very special version of the truth. He was also a highly experienced outbacker, with a sharp eye for seeking out very old rocks and their very odd stories. As John would say, Australia is both the oldest and the oddest of all the continents on planet Earth. Old because it contains rocks approaching 4 billion years in age. And odd because these rocks have hardly moved up or down over the past 2 billion years or so. Australia is the slumbering continent. At the outset, therefore, my expedition with this particular Apollo Missionary sounded like the calmest and most relaxing of undertakings. But that turned out to be a foolish assumption.

John and I had planned to drive across northern Australia in a wide arc from the mining town of Mount Isa in the east, towards the hot and humid town of Darwin in the west, via the hinterlands of the Gulf of Carpentaria. Our aim was to gather information about the early evolution of the eukaryote cell, and its context in Deep Time, most especially from rocks in this region that were laid down during a period which I had already started to call the 'Boring Billion'. Not boring in the sense of uninteresting—that was a kind of ironic jest. But boring in the sense of its curious chemical fingerprint—a vast and seemingly improbable interlude of very stable carbon isotopes that had lasted from about two to 1 billion years before the present. Indeed, the Boring Billion seemed almost as improbable as Australia herself.

Our expedition was to become memorable owing not only to these marvellous rocks but also to the kindness of the Australian Geological Survey Organization—once affectionately known as AGSO—whose staff had recently settled down into new buildings in Canberra. By way of explanation of what follows, it has to be admitted that AGSO was becoming a rather 'interesting' institution at the time of our expedition in the year of 1998. Everything seemed to be in place for its meteoric development. I say 'meteoric' here because, just like a meteor coming in to land from space, its progress was about to be downward.

I sensed a feeling of vertigo almost as soon as I entered the shiny new building of AGSO, with its steel and glass frames surrounded by groves of gum trees set between wide green lawns. While working away in the inner sanctum of the library, I became aware of a strange hissing noise. Ploughing my way through stacks of books and pamphlets like an old bushwacker, I came unexpectedly across a mock waterfall, replete with rocks and palm trees. These faux features, it seemed, were there to help keep books in the bookstore damp but they also had a less convivial effect as well. They induced an uneasy feeling of superficiality and spin that was at risk of permeating the whole building. According to John my host, I had arrived in Canberra at a 'difficult time'. Explorations into Deep Time and the history of life itself, in which Australians had once excelled, were in the process of being squeezed to death by a philosophy called Economic Rationalism.

'Economic Rationialism' was a polite name for that phenomenon otherwise known as Reaganism in the USA and Thatcherism in the UK. Arguments in favour of this form of economics sounded rational enough. As I understand it, those arguments have tended to run broadly as follows. The money supply of a modern economy should be carefully rationed. And the whole of society should be made to compete for these funds by offering ever greater efficiency savings. Those at the top of the pyramid should then be allowed to pay themselves huge bonuses from these savings, because they are motivated to work harder by more pay. And those at the bottom of the pyramid should be threatened with unemployment, because lesser people are motivated to work by offers of lesser pay. Our rewards should come, therefore, by means of a process of trickledown from the top.

Here, then, was a fine example of Topdownian thinking. To suggest how such trickledown economics could really work in practice, fingers were pointing hopefully towards entrepreneurs who were flourishing across North America and England, as well as here in Australia. As might be expected at such a time of financial crisis, the offices of AGSO were full of geologists and geophysicists gathering in huddles. They were not full of palaeontologists in huddles because those time-travellers had already been sent on their way. Historical studies of the biosphere were believed to have rather little relevance to an organization steering towards wholly material goals.[2]

Observing the offices of the AGSO that year therefore felt deeply unsettling. It was a bit like peering down upon a dying coral reef through the hull of a glass bottomed boat. This cruel analogy seemed somewhat apt at the time because the very walls of its new building were constructed from vast panes of glass. These see-through walls were intended to discourage workers from falling asleep at their desks, said some. Or from planning clandestine meetings, said others. But the effect on many was rather the opposite—it invoked a feeling of paranoia. It was a mournful thing to observe a great institution in such a state of turmoil. Part of the problem, it seems, was this: a storm of government directives had been followed up by a series of management initiatives that, in turn, gave rise to an exchange of toxic invectives between the staff themselves. This chain of blame was slowly poisoning the once great Geological Survey of Australia. The study of life in Deep Time in Australia was therefore to come very close to extinction itself.

That, at least, is how it seemed to me just before our expedition. But a trio of factors made the journey more than usually memorable, too. In the first place, the Land Rover Explorer—with its long wheel base and dusty canvas awning—awaiting us up north was reputedly in poor physical shape after endless rounds of spending cuts. Our transport was therefore at risk of succumbing to the ravages of gravel roads and eventually to gravity itself as we rammed our way through thorn trees, burnt stakes, and tussock grass towards the O'Shannassy River. Secondly, there was to be no extra field support—it was to be just John and myself plus any bits of kit we might be able to glean from a derelict field base on the edge Mount Isa. And finally, there was to be no provision of two-way radios for safety. If we ran out of water or got bitten by spiders, we were expected to suffer in silence. Unhappily for John, silence did not come easily.

A brush with destiny

The padlocked compound at Mount Isa looked like a scene from an outback disaster movie. Emptied of staff, its cheap plywood doors were broken down and its floors were strewn with old mattresses, as well as dented pots and pans and defunct gas stoves. From this debris, we gingerly selected two filthy old swag bags.

This term 'swag' has seemingly been lifted from an old diggers song, like 'Waltzing Matilda'. As a term, it conjures images of bristly old bushwackers replete with their cow-hide hats, dangling corks, and tucker bags, snuggling down to sleep in a billabong. But in theory, the swag bag is a brilliant way of sleeping under the stars when living out in the outback. It consists of a number of separate parts. First, there is a huge rubbery tarpaulin to keep out damp and deter those less-than-friendly Australian snakes and spiders. Happily, the rubbery smell of this tarpaulin—called a 'tarp'—is enough to send most kinds of vermin scuttling. But there is more. Wrapped inside the tarpaulin lurks a bendy foam mattress together with a couple of foam-filled pillows, usually covered in dust and dead cattle ticks. Needless to say, after a week of sleeping in such a swag, its combination of odours could deter most wildlife in the vicinity. As we spluttered out of town and headed for the Gulf of Carpentaria, our ageing Land Rover looked more like a tinker's truck than a serious scientific expedition, loaded down with swags and cans, pots and pans. On the morning of departure the weather turned cool and gusty. So we set off rather gloomily, concerned that the roads ahead might be washed away and that our beloved fossils would escape capture. Dirt roads in the outback can make for a delightful drive if the weather stays dry, which it will do for many months at a time. But even a modest fall of rain can turn the dusty soils round here into a sea of glutinous red mud. Our greatest fear, therefore, was to be caught by a massive downpour of rain and then be stranded for days in the wilderness.[3]

Under leaden skies, we scurried along the roads and mining tracks. As we took turns at the wheel, we talked about our first mission—to explore claims for a huge meteorite impact crater some 200 kilometres to the north of Mount Isa. The American astronomer Eugene Shoemaker—whose remains now lie on the moon—had discovered a site around Lawn Hill within the Northern Territories while exploring a series of satellite images. Our job was to test this evidence on the ground and then to report back to a newly formed mining company, based at the Century Lead-Zinc Mine. Was this truly an ancient and important meteorite impact crater? Or was it just an oddly shaped hill, as they hoped? If the former, then their quarrying for ores could have its own unwanted impact upon the landscape.

I shall never forget the scene that greeted us as we entered the gated compound of the Century Mine near Lawn Hill (Figure 31). It was like a scene from a corny old James Bond film. The mining compound,

Figure 31. Our Land Rover is parked near to the rim of the 20km-wide Lawn Hill crater, here seen as a ridge of rocks in the distance. This thorn tree was filled with a flock of chirping budgerigars.

remember, was set inside a huge depression some 20 kilometres across, and surrounded by high wire fences with DANGER KEEP OUT signs. At the gate, we were obliged to hand over camera equipment and to wear company clothing. Looking around, I could see people marching about in buff uniforms, looking like Bond-movie extras. Anytime soon, we joked, the central hub of the 'crater' would roll back to reveal a battery of nuclear missiles, all pointed in the direction of Shepperton Studios in England. But we never came across either the Bond villain or even his white Persian cat. To be strictly honest, we were royally looked after during our stay.

Hipparchus

John was highly delighted by this great circular structure. It had all the features of a meteorite crater—a central region showing smashed up basement rocks, surrounded by a broad circular rim filled with slumped and broken seafloor sediments. He began talking freely about his old lunar exploits. He, or rather we, began walking about freely like Apollo Mission astronauts as well, sampling the bright white rocks, delving into the dark shadows, and throwing up clouds of red dust.[4] The Lawn Hill meteorite, we concluded, had crashed into a tropical ocean with great force, crushing and flattening a disc-shaped region at its centre while sculpting out a circular ditch around its edge. And it was into this ditch that tropical limestones had slumped while still pliable—to form huge swirls of sediment like a mint fondue. But quite when this had happened, we were not at all sure.[5]

Here was a crater almost exactly like those craters on the Moon. But how had those been formed? The answer was first indicated by none other than Robert Hooke in 1665. When leafing through that marvellous tome of *Micrographia*, it comes as a shock to stumble across his final set of experiments. Not only was Hooke the first to draw and then describe a lunar crater in great detail, he was also the first to experiment on ways in which such craters might have formed.[6]

For Hooke, it all began after pointing a 10-metre telescope at a star-filled sky in the October of 1664. During one such evening, he made some meticulous drawings of the crater Hipparchus, close to the equator of the moon.[7] In his excited and breathless fashion, Hooke tells us that this crater 'appears a very spacious Vale, incompassed with a ridge of Hills, not very high in comparison of many others in the Moon, nor yet very steep.' Most charmingly, Hooke describes Hipparchus as though standing alongside us inside the Lawn Hill crater itself:

> ...from several appearances of it, seems to be some very fruitful place, that is, to have its surface all covered over with some kinds of vegetable substances; for in all positions of the light on it, it seems to give a much fainter reflection then the more barren tops of the incompassing Hills, and those are much fainter than diverse other cragged, chalky, or rocky Mountains of the Moon. So that I am not unapt to think, that the Vale may have Vegetables *analogous* to our Grass, Shrubs, and Trees; and most of these incompassing Hills may be covered with so thin a vegetable Coat, as we may observe the Hills with us to be, such as the short Sheep pasture which covers the Hills of *Salisbury* Plains.[8]

In truth, there is no vegetation on the moon, of course, but there was plenty all about us here at Lawn Hill—craggy chalk hills surrounding a vale carpeted with short grass and stands of shrubs and trees, with flocks of budgerigars swooping across the crater vale. But what really stands out is the manner in which Robert Hooke set out to experiment with ways in which a lunar crater might be able to form. For one experiment, he filled a bowl with white clay—of the kind then being used to make tobacco pipes—into which he then dropped a series of heavy objects. But Hooke could not easily imagine how huge objects could be made to travel through space and then land on the Moon nor, indeed, how its surface could be made to react so sloppily. For a second experiment, he therefore boiled a mixture of alabaster and water in a basin and produced ring-shaped bubbles which he likened to volcanic processes.[9]

While Robert Hooke had opted for the volcanic explanation of lunar craters, the Apollo astronauts had firmly confirmed the first experiment—they are indeed the result of impacts from meteorites arriving from space. And this is exactly the conclusion we came to with the Lawn Hill crater, too. It had been formed by a huge body—at least a kilometre across—colliding with the surface of our planet. But when exactly had this Lawn Hill catastrophe happened?

A sticky problem

We attempted to trace out the tropical limestones that infilled the crater rims, driving our Land Rover away from the crater floor and off the gravel roads and then far into the Australian bush. At one point, our route took us along a mineral prospector's trail—a faint path through dusty thorn trees that had suffered decades of neglect. To keep on track, we therefore followed a sequence of faded pink ribbons that had been teasingly tied to the thorn bushes every 50 metres or so by some previous visitors. Our progress was far from straightforward. At one point, after 10 kilometres of bushwacking, we lost sight of the pink ribbons and found ourselves back at our starting point. At another point, we drove into the deep gulley of a creek bed and had to haul ourselves out with spades and winches.

We arrived at our camping site just before sundown, set down by the O'Shannassay River and some 20 kilometres off the dirt road. Unloading the wagon was, as always, an anxious journey of discovery. Things would appear that we had never packed—like an old pair of trainers. Others would disappear forever—like our cooking stove. This minor calamity had forced us to cook our dinner over a fire of gum tree logs. A rare and tender joint of beef steak was on the menu that night, cooked in John's inimitable outback style. After we ate, our reeking old swags were placed well away from the dining area and set out beneath an old fig tree, itself beside the babbling waters of the O'Shannassay River. No snakes or spiders would bother us there, we felt. No saltwater crocodiles either—just a tinnitus-like chorus from the crickets sitting beneath a full moon.

To avoid the worst of the heat next day, we began sampling the nearby hillside soon after dawn. With measuring stick plus an abney level—to calculate the thickness of the rock beds—and a gamma ray logger—to read off the amounts of clay from radioactive signals—we climbed through layer after layer of old seafloor. Piece by piece, the story-book adventures of a very old Cambrian world began to emerge in our minds. We found superbly preserved molluscs and tubeworms in layers resembling ancient tidal flats. Above them sat a thick stack of submarine sand ridges, formed almost entirely of fossil skeletons of echinoderms—the earliest ancestors of sea urchins and starfish.

Scattered over the rocks hereabouts lay the knappings of stone age hunters. One could visualize where each craftsman had sat on the rocks, and how each piece of chert had been selected and then dressed. But it was at the top of the hill—in the youngest layer of sediments—that a real treat awaited. Layer upon layer of wrinkles and domes were laid out on the ground, looking here like a monumental flaky pastry. These were cousins of those strange markings that Darwin had found during his *Beagle* voyage; that James Hall had puzzled over near New York; and that Tyler had later seen around Lake Superior: stromatolites galore.

The Heartbreak Hotel

According to our studies, the arrival of the Lawn Hill meteorite had broadly coincided with some of the very first reef extinctions in the fossil record—the destruction of symbioses between sponges and algae near the end of the early Cambrian.[10] The Lawn Hill meteorite crater was, however, a mere prelude to our investigations into early life around the Gulf of Carpentaria. Our plan was to push back a further 1000 million years. Back, indeed, to the interval when the earliest eukaryotic cells had been evolving—during the period of time known as the Middle Proterozoic.

This plan meant traversing some of the flattest and most empty terrain I have ever seen—called The Barkly Tableland. It is so flat here that it is easy to see the gentle curvature of the Earth. And so empty that we never spotted a soul during hundreds of kilometres of driving. Even while crossing the steppes of Mongolia, I had at least seen horsemen; or in the

Arabian desert I had at least met with Bedouin. But here, for some 600 kilometres or so, we drove in almost complete silence through a savannah landscape crossed by a single track road with barely a swerve.

After a particularly long road-eating stretch, we were relieved to spot a line of scrubby trees in the distance, floating above a watery mirage. Little by little, the grass grew greener, the trees grew taller, and the hills got hillier. We had finally entered a region drained by the MacArthur River—and home of the infamous Heartbreak Hotel.[11]

No hostelry could have repaired our thirst quite as well as did the Heartbreak Hotel. That was lucky, because it was the only hotel in the area. Purple bourgainvillias draped its verandas while cockatoos croaked around the shower blocks of this remote and peaceful retreat. Once we had unpacked our Land Rover, our aim was to relax, to put our feet up, and to drink a beer or two, with the clear aim of drinking up the tranquillity all around us. The noise and the dust of the road was now far behind, we told ourselves. It would become but a distant memory. Or so went the plan. But this plan—like many of John's plans—was not to last. The hotelier was a man called Kevin. And Kevin, we soon learned, was rather keen on helicopters.

Helicopters are a delightful way to see the world in theory, but less so in practice. Many a geologist has therefore developed a reflexive dislike of helicopters. But being optimists by nature, we ignored our inner voices and hopped aboard his flying machine.

Soon enough, we were sailing southwards from the Heartbreak Hotel and hovering over a fantasy land of pillars and columns sculpted out of red sandstone in the nearby Abner Ranges—the 'Lost City'.[12] Within ten minutes, we had landed in the middle of a highly fractal landscape that few have ever entered, sculpted out of soft silica sandstones known to be about 1450 million years old—almost three times as old as the Cambrian (Plate 13).[13] Aboriginal artwork, caverns, flowers, and birds were scattered around us like gifts. It was like a little piece of paradise.

Eons of erosion and cracking had carved the bedrock into something resembling a lost city replete with tower blocks and cathedrals, with intervening areas resembling arenas and piazzas. But to the trained eye of a geologist, other strange features could be seen in these rocks, too. Each layer showed the remains of large ripples that had evidently swirled this way and that on the sea floor, as though from the effect of tides on an ancient beach.

Tidal sediments can be taken to imply tidal forces. And tidal forces can be taken to imply the presence of the Moon. Here then was evidence, if anyone would ever have doubted it, that the Earth–Moon system was in place long ago, and certainly as far back as 1450 million years past.[14] So it felt delightful to stand among these tidal beds beneath a tropical sun, and then look up and see a brightly lit Moon still sailing serenely across the deep blue sky of the outback. Perhaps the most marvellous thing about these 1450 million year sediments, though, is the fossils they have been found to contain—known as the Roper Group fossils.[15]

Emmanuelle Javaux from Liège in Belgium, together with Andy Knoll from Harvard and Malcolm Walter from Sydney in Australia, has reaped a rich harvest of single-celled organisms from these rocks.[16] Some are of a size so large (several tenths of a millimetre across) that they have been regarded as the remains of creatures much more complex than bacteria. Particularly telling here are the details shown by some of these Roper Group fossils. Some looked like squashed raisins, with arc-shaped wrinkles from having been pressed beneath layers of sediment. These are called *Leiosphaeridia*.[17] Yet others, called *Tappania* after American micropalaeontologist Helen Tappan—had spines poking out of a central ball, rather like a little green conker (Figure 32). As with modern plankton, these spines may have helped the little organism to float in the water column, and stay closer to the sunlight (Plate 14b).[18] A few even showed highly intricate patterns over their walls, looking as though they had been covered in Lilleputian bubble-wrap.[19]

These conkers and bubble-wraps may sound unremarkable, but they are truly rather important. That is because their spines and intricate surface patterns have been taken as evidence for the presence of a microtubule

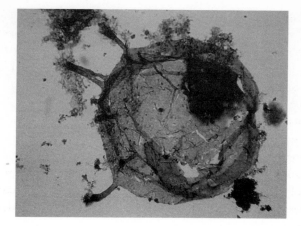

Figure 32. This conker-like microfossil, called *Tappania*, was secreted by one of the earliest known complex cells, some 1450 million years ago. It was probably some kind of Green algal resting stage, according to the ideas of Malgorzata Moczydlowska 2011. Image kindly provided by Emmanuelle Javaux from the Roper Group of Australia. The fossil is about 120 microns wide.

cytoskeleton—a feature unique to eukaryote cells.[20] Forams, algae, plants, and even our own cells make use of a cytoskeleton to help define its shape, or change it during growth and reproduction. If so, then these fossils from the Roper Group represent the earliest acceptable remains of such 'modern' cells in the whole of the fossil record.

Now there are some older candidates that some regard as respectable eukaryote fossils. One is a little bag-shaped structure called *Valeria*. It is about 1600 million years old. But it lacks any signs of having had a microtubule skeleton—it has neither spines nor pimples across its cell surface, so it fails the test. And there are younger fossils that we may accept more readily. Most famous of these is a tubular microfossil called *Bangiomorpha* (Figure 33). This pretty little fossil from arctic Canada may be as old as 1270 million years.[21] Interestingly, it shows evidence for cell division both around and along its tiny cylindrical frond, in the manner of the living Red alga *Bangia*, which is also a tiny filamentous form. This suggests that

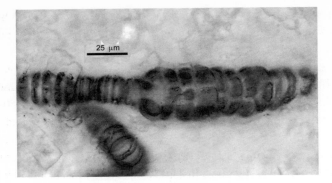

Figure 33. These filaments of *Bangiomorpha*, embedded within cherts about 1270 million years old in arctic Canada, show patterns of cell division, and cell specialization. As the name implies, they invite comparison with patterns found within the living Red alga called *Bangia*.

Note. Image kindly provided by Nick Butterfield.

Red algal seaweeds had begun their adventurous journey into multicellularity by that time.[22]

By 1000 million years ago, it can therefore be argued that the complexity of life had taken a great leap forward, from bacteria towards the forms of Green algae and Red algae.[23] By this time, there is good evidence for comples cells living on land as well, preserved within the Torridonian rocks of Scotland (Plate 15).[24]

Amelia

Panting heavily in the shade of a black thorn tree, John Lindsay was scanning the horizon for the next staging post—another tree, some 500 metres beyond. I could hear my own heart thumping rapidly from the heat of the afternoon. We had been travelling in stages across the sun-parched scrub for four hours and were beginning to wonder if we would ever make it back to our old Land Rover, still far away at the head of Kilgour Gorge. John and I had foolishly fallen into one of the deadliest traps known in the Australian outback. We had only intended an hour long walk from the vehicle. But we ended up venturing far into the bush, without food or sufficient water.

The morning had begun well enough with a light fresh breeze. We had travelled some 75 kilometres from the Heartbreak Hotel, among winding dirt tracks through open scrubland, seeing barely a single moving creature. Leaving our increasingly decrepit vehicle behind in a small clearing, we were intent on searching for those signs of ancient life that can reputedly be found on the far bank of Kilgour Gorge, a few miles to our west, laid down in a great cliff of red rock formed by the charmingly named Amelia Dolomite (Plate 14a). It was from these 1600 million year old beds that the palaeontologist Marjorie Muir had described an assemblage of fossils that seemed wholly bacterial—just like the Gunflint chert microfossils of Barghoorn and Tyler. And like the Gunflint fossils, they came without a sniff of evidence for Green, Red, or Brown algae.

The floor of the Kilgour river bed held a green lake of standing water, temptingly shaded by large palm trees that beckoned us down into the idyll. Once down in the gorge, however, we found that the water holes were much too full to wade through—freshwater crocodiles permitting—or even to walk around. We were therefore forced to quit the shade and to climb up on to the arid flanks of the gorge, fighting our way through waist-high tussocks of spiky spinifex grass. Hot and stinging, we arrived at the next bend in the gorge where, after several attempts, we were able to find a path along the dry river bed that meandered onwards for several kilometres. By the time we reached the other side of the gorge we were both tired and parched. Tormented by flies, we climbed up through the scrub to hammer out the rocks and fill our backpacks full with lumps of stromatolite and dusty Amelia dolomite.

It was in associated cherty layers that Marjorie Muir had discovered her microfossils. These forms were very like those from the Gunflint chert, nearly 200 million years older: long filaments and rounded spheres of bacterial cells, including forms allied to cyanobacteria like *Oscillatoria*.[25]

Many people have hunted hard through these rocks. But they contain no clear traces of eukaryotic life.

Danger awaited us, though, after a lunchtime spent working without lunch. Climbing up the red dolomite cliffs, John developed a serious pain in his right knee, making it hard to descend the way we had come. It was then we realized we would not be able to scramble back down into Kilgour Gorge at all. Instead, we were obliged to hike around the river bed in a wide arc, keeping ourselves at the level of a layer of gritty rock called the Tatoola Sandstone—and thus avoiding the heads of creeks and gulleys that fed downwards into the limestone gorge below. The map showed us that, if we circled carefully around the creeks, it would be a hike of about 20 kilometres. It was a hot, cloudless afternoon. Looking at our watches, we reckoned this trek would take about five hours and, all being well, we should be able to get back to our vehicle just before sundown. And all without water.

Our minds soon became tired and befuddled from the heat and the utter lack of breeze, as we trudged across the grasslands from one isolated tree to the next. The sandstone ridge on which we found ourselves supported little vegetation except for spinifex grass, dotted here and there with the mounds of termites, held together with saliva cloyed with parched red dust. In no time at all, our own saliva felt cloyed with red dust as well.

To keep moving, we set ourselves a target. We would walk for some ten minutes across the blistering plateau, and then follow this with five minutes of panting in the shade of a tree. We even gave up talking—a rare concession for John—to conserve moisture. Although the ground was scattered with flint implements, neither of us cared to bend down and examine them in case we blacked-out. After more than five hours of this torment, the sun began to sink, bringing with it a warm breeze, and then a slightly cooler breeze. Then at last, we espied in the distance the tiny white speck of our Land Rover, our passport home.

At sundown, we squeezed thankfully beneath its oily chassis to let fluid from its water tank trickle over us. We laughed with sheer relief, for it had felt like a close call. Next day in the shower, I found I was host to five red cattle ticks—the symbionts were striking back.

Emmerugga

After our mad-dog experience with the Amelia Dolomite, our next horizon was selected with more common sense. Set right beside the road, the Emmerugga Dolomite was easy of access. And it presented some of the finest stromatolite fossils to be seen anywhere in the world.

Those Cambrian stromatolites we had met in the fossil beds by the O'Shannassay, remember, were mere wimpish wrinkles—like a tray of flaky pastry. They could be forgiven for their wimpishness because they were, of course, a mere 520 million years old; faint echoes in the long saga of stromatolites. Even these stromatolites in the Emmerugga dolomite were wrinkles too, but they were not wimpish. They were up to a metre tall—giants of their kind—and strangely formed like conical witches' hats, so that they were called *Conophyton*, meaning 'conical plant fossil' (Figure 34).[26]

Conophyton has a particular claim to fame in our story. This kind of conical stromatolite has never been found in rocks as young as the Cambrian, when complex animals first exploded on to the stage some 542 million years ago. But they abound in much older rocks, especially those of about this age, some 1650 million years ago. According to my colleague John Grotzinger of Caltech in California, they tended to grow on seafloors in deep water, close perhaps to the zone where sunlight could only just penetrate.[27] In the modern tropics, that would mean these waters were very clear indeed, and in the region of 100 metres or so deep. In this way, it has been suggested, those poor light-loving cyanobacterial cells would crowd together hopefully onto the very tops of any slight irregularity, causing it to grow upwards into a cone-like shape. In other words, these fossil witches' hats suggested the presence of microbes that relied on photosynthesis.

Such clear waters were starting to sound a bit familiar. I had met with deep and clear waters like this before—in the Sargasso Sea, and again

Figure 34. NASA geologist John Lindsay measuring the height above the ancient seabed reached by a 1650 million year old cone-shaped stromatolite called *Conophyton*, found within the Emmerugga dolomite of Queensland Australia.

around the edges of the Pedro Bank. On its own, this image wouldn't be able to tell us very much. But as we shall see, it may help us to put together the pieces of a much bigger picture.

The idea that was forming in our heads at that time was as follows. We had stumbled on an early world in which the top layers of the ocean had an almost gem-like quality, both remarkably clear and stunningly blue, reaching down as far as light could possibly reach on the seafloor. Maybe the ocean was like a modern Blue Water ecosystem but spread out around the globe. Could that really be true and how on Earth could we hope to test such a hypothesis?

The not-so-Boring Billion

Testing this concept of an unusually calm and blue world ocean was an important goal during our expedition. Throughout the field season, we collected samples of limestone at regular intervals through any outcrops that we could reach—and even those we could not reach, by sampling those drill cores stored in mining sheds.

It has to be said that John had many strange skills, involving the collecting of Moon rocks, Australian maps, Pre-columbian pots, and even Jacobean pans. But his greatest wizardry was tracking and then negotiating access to hidden core sheds and their secret cores. Unglamorous work. Physically demanding too, because heavy core trays are protected with jagged steel corners, seemingly designed to crush and tear at fingers and thumbs. But the core sheds of a mining company or geological survey are not really intended as places of torture. Rather, they deserve comparison with the inner sanctum of a temple. In them are enshrined those chapters and pages rescued from the geological bible—many kilometres of core, drilled from rocks deep down beneath the MacArthur and Mount Isa regions. These rocks we were about to examine, remember, had been brought blinking into the Australian daylight, stored in trays and neatly stacked in rows of racks.

Bit by bit, a huge succession of chalky rocks—limestone and dolomites—was therefore assembled and brought to Oxford for chemical analysis, especially of its isotopes, and most especially for analysis of the element carbon. Now every chemical element, including carbon, has its own distinctive character. There are many to choose from, but life itself is very particular in its choice. Life prefers its elements to be light and nimble rather than heavy and ponderous. That is why life likes the flighty elements of hydrogen and oxygen and abjures the toxically heavy elements of lead and gold. Indeed, it could be argued life might be rather difficult without ready access to about eighteen of these chemical elements. Central to nearly all of life, though, is the element carbon.

One hears the name of 'carbon' quite often in the news—carbon dioxide and carbon emissions; carbon trading and carbon budgets. But not all carbon atoms are quite the same. Some weigh in at an atomic weight of twelve, while others tip the scales at a massive fourteen. A third kind weighs in at an intermediate thirteen. These different kinds are called 'isotopes', and they vary in atomic weight because of the different number of neutrons in the nucleus. Now living organisms take these weights very seriously indeed. Your own body shows a much greater concentration of light carbon 12 than is found in the air or the sea outside. That is because your body is basically built from proteins gathered from plants such as potatoes, or from plant-eating animals like sheep. And the chloroplasts of plants show a strong preference for picking up these lighter carbon isotopes when they are building molecules of glucose from carbon dioxide during photosynthesis. In other words, we can tell about the activities of chloroplasts in the past by looking at the isotopes of carbon in rocks.

Back in Oxford, we had a machine, called a mass spectrometer, that could smash the carbon isotopes out of these rocks, and separate them by weight.[28] Feeding our rocks into this mass spectrometer, we began to notice an extraordinary pattern. We were looking, remember, for changes in the ratio between carbon 12 and carbon 13. Today, these ratios fluctuate regularly at all manner of scales. In the shell of a seaside clam, they vary from day to night because of the effect of light on the net amounts of photosynthesis and respiration, and they change from season to season for rather similar reasons. If we look at Foraminifera in deep sea cores, these values vary on a scale of thousands of years, because of the effects of ice ages upon warmth and light. Much more rarely, we can find really huge wiggles, as indeed at those times of mass extinction we have recently explored, when the levels of photosynthesis seemingly collapsed dramatically.

So we were used to carbon anomalies—carbon ups and carbon downs—not only today but during the whole of animal evolution from the Cambrian period onwards. But as we traced these wiggles backwards in time, they at first got bigger and bigger, especially during the Snowball glaciations, some 730 to 630 million years ago. We even found carbon isotope anomalies during a much earlier period of Snowball glaciations, from 2400 to 2200 million years ago.

But the strangest thing emerged between these two peaks. From about 2000 to 1000 million years ago none of this wiggliness seemed to be present in the rock record. Nothing much happened to the carbon signal, even in the very best preserved of sediments such as those from around the Heartbreak Hotel. The curve seemed to be as flat as the Barkly Tableland (see Figure 1 in the Preface).

A name for this curious interval was suggested to me during a lengthy interview with a journalist. There was, it seems, a media term for very dull and wealthy people. They were called Boring Billionaires.[29] The interviewer suggested that we might call this interval the Boring Billion. And that is exactly what we did during an international conference in Adelaide in 1998.[30] Needless to say, the idea was not well received at first. The physicist Paul Davies said he was highly sceptical. But the concept has survived the test of time. Indeed, it now has its own Wikipedia entry.

Dangerous liaisons

Our initial explanation for the Boring Billion went something like this. From 2 billion to 1 billion years ago, the world experienced little or no major continental collisions or ice ages. Without continental collisions, the resulting rapid mountain building, and consequent rapid erosion and deposition of sediment in the sea, supplies of nutrients such as phosphorus to the oceans were therefore scant. Without nutrients, the oceans could not form great blooms of algae. Hence the surface of the ocean was characterized by something that resembled a vast Blue Water ecosystem.

Other scientists have since explored yet other factors that may have been at play here. One of these, suggested by Don Canfield in Denmark, has invoked a shortage of sulphate ions in seawater—the so-called Canfield ocean effect.[31] Another suggested by Andy Knoll of Harvard and his colleagues has involved a dearth of the element molybdenum.[32] But, however we look at it, we can say that the Boring Billion was an interval during which nutrients were scarce at the surface of the oceans. And that means that their internal recycling could have been of paramount importance. This is a puzzle to which we shall need to return.

Sorry tail

There is now a sting to the end of this tale. John Lindsay had been pushing through his own agenda for good science, but bureaucrats at the Geological Survey in Canberra saw things differently. Years of squabbling ended up with the worst of all possible solutions—John was fired from his post. The Boring Billion research therefore marked the beginning of the end for him. In one terrible week in July 1999, John was given his marching orders, and his wife Kay—a life scientist—was diagnosed with terminal cancer. The decline of his family life was a sad thing to witness. By 2001, Kay had succumbed to the disease and John had returned to NASA at Houston, where he eked out a precarious existence on early life and Apollo Mission projects, through the kindness of David McKay. While he was away, his house in Canberra was entirely burned down by a firestorm. By 2007, John was also fighting cancer, and he finally lost the fight on 20 June 2008. By coincidence, he died on the very same night that I sat by my father's own deathbed.

Then in 2010, I met one of John's colleagues in Canberra. Our pioneer work on the Boring Billion had turned up trumps, providing a baseline against which to measure the numerous highly valuable ore deposits of this age and region. John's intuitions had been right on the ball.

CHAPTER 11

GAIRLOCH

From heaven to hell

We now come to one of the most baffling mysteries presented to us by the fossil record. The rocks suggest that a major change has taken place in the mode and tempo of biosphere evolution through the course of Deep Time. We can see this best when we look at rocks from 2000 to 1000 million years ago, during what I have called the Boring Billion (Plates 13–15). Fossil species during this epoch appear to have survived unhindered for huge spans—up to hundreds of millions of years.[1] Nor can we find much evidence here for cycles of ecosystem boom and bust. Instead, we find a long period of quiescence with barely any chemical oscillations in the oceans, and no big freezes to speak of. This may sound boring to a human geochemist; but it was arguably Microbial Heaven.

When we fast-forward towards a younger world up to about 600 million years ago, we find a different pattern—with plentiful booms and busts taking place in ecosystems across the planet. Species seem to come and go from the fossil record at a much more hurried pace after this time. But then, about every 30 million years or so, we discover evidence in the rocks for huge bust-ups—the mass extinction events—when numerous species died out around the globe in relatively short spans of time. These oscillations first became prominent after the so-called Cambrian Explosion, some 542 million years ago, when the modern animal groups appeared. It is from this time onwards that Earth began moving away from Microbial Heaven towards Animal Hell.

Now some old scientists, such as Lyell and Darwin perhaps, might have tried to explain away these paradoxical contrasts by suggesting that fossil preservation gets worse as we travel backwards in time. In other words, that the Precambrian fossil record is too feeble to reveal any mass extinction events. But as I have argued in *Darwin's Lost World*, exactly the opposite seems to be the case. The quality of the fossil record seems to get *better* as we travel back in time. Before the advent of animals and predation, cells were not always chewed to death. They could enter the fossil record almost like flies trapped in amber. If so, then we ought to have confidence in what the fossil record is trying to tell us here. And it is arguably saying that mass extinctions are—in geological terms—a relatively recent phenomenon. They may have been lacking from much of the early history of life.

Armed in this way, we can begin to perceive some major changes in patterns of symbiosis through time as well. It was during the Boring Billion, remember, that free living cyanobacteria are thought to have been enticed into the cells of some foram-like ancestor. For some reason, these lodgers stayed so long that they became permanently embedded, allowing themselves to be turned into those structures we now call the mitochondrion and the chloroplast. This imprisonment was most likely assisted by the capture of some of their genetic information by the host, making the escape of its guests almost impossible, like the impounding of passports by a distrustful landlord. All forms of algae—and hence all forms of plant—are thought to have arisen in this curious way, and arguably at this time too, by a succession of symbiotic partnerships. Likewise, all the organelles of our own cells, such as the nucleus and the mitochondrion were formed in this way, and about that time. At the biological level, many of the bits and pieces in our cells have thereby combined together, from different ancestors, to make a eukaryote cell. An ancient symbiosis was able to turn into a completely new kind of individual. It is an astonishing thought, and it opens up all sorts of new and difficult questions too. Could symbiosis be regarded as the norm for the evolution of complexity, from the origins of life all the way towards complex colonies, such as those of ants and humans? Do the old communities of the past often become the new individuals of the future?[2] This is a fascinating area but, alas, we shall have to put it aside for consideration in a further book on the origins of life itself.

Returning to our living coral reefs, we have seen the evidence that can be gathered for symbiotic trysts, such as those between corals and algae, or protozoans and algae, within their Blue Water ecosystems. Some of these have taken their habit of symbiosis to the point of obsession. We have also seen how these partnerships are currently breaking apart, largely owing to man-made environmental stresses. These hosts are therefore expelling their symbionts, and turning inside-out again. Nor is the death on the reefs the first time that such liaisons have been broken apart. The fossil record of such partnerships now reveals some ten or so cycles of boom-followed-by-bust. Algal guests, it seems, have seldom stayed welcome in their hosts for longer than 30 million years. After that, they have seemingly been expelled at times of wider ecosystem collapse, during those mass extinction events.

If this is a correct reading of the fossil record, it implies that no new kinds of organelle, and no wholly new kinds of algae, have been permitted to arise by symbiosis during the last 540 million years or so. Only in the preceding eons does the forging of wholly new organelles by symbiosis seem to have been possible.

This all sounds very puzzling. So how might it all be explained? A successful hypothesis should be able to predict not only the patterns of energy flow within a living ecosystem but also within a fossil system too. Modelling of that kind is clearly going to be a complex undertaking, and at the present it is best regarded as an agenda for future research. But as with Darwin and his homing-pigeons, we may be permitted to make some progress here by exploring simple models, parallel systems and parables.

The twin pyramids

In 2001, as the horror of the 9/11 tragedy broke in the news, my elder son Matthew and I had been at work on the mechanisms behind ancient massive loss of life—natural tragedies that lacked the deliberate brutality that only conscious human acts can achieve. We had been working on the collapse of food pyramids during mass extinction events. Or to be more precise, Matthew had been working on it alone and I had been giving him 'advice'. Of especial help here were his skills as a computer programmer working with JAVA code, and his experience with modelling neural networks.[3] At first, we decided to build a model ecosystem a bit like the Eiffel Tower in Paris—a pyramid-shaped structure with a wide base and a pointed top. The apex would stand for some organism at the top of a food pyramid, say a top predator. Beneath it was positioned a lower feeding level with three nodes—to stand for three organisms on which the top predator might have fed, including smaller carnivores. Lower down there was a level with four, and yet another with five, nodes, for lower parts of the food web, including insectivores and herbivores.

Each of the nodes in our model was allowed to behave like a light switch in an electric circuit. It could be switched on or switched off, but it could also be left partially open so as to act as a kind of filter. And rather like an electric circuit, too, these nodes could then be linked in a wide variety of interesting ways (Figure 35 and Plate 16) allowing us to test the various effect of different types and levels of interconnection. Some of the food pyramids had lots of cross links (Figure 35 at left) while others had only a few, and one end member had only vertical links (Figure 35 at right).

Next came the tricky bit. A rapid succession of random numbers was fed into the various pyramids via the nodes at the bottom level. This was to simulate environmental information coming into the food chain at the base of the food pyramid—such as changes in light, or nutrients such as phosphorus. Each new input value differed little from the previous one but, being a random series, there were occasional blips in which large changes in value could take place. These blips helped to simulate environmental perturbations, such as those that might be brought about by storms, hard winters, and plagues.

It is important to emphasize that our pattern of inputs to this system was always similar—just a set of random numbers. But the ways in which these variations ramified through the pyramid turned out to differ drastically and, to our eyes, in rather unexpected ways that seemed very puzzling indeed. Of most interest was the manner in which the paramount creature—the top predator, say—could be affected by all these ups and downs. An output from this top node could be printed out or displayed on a screen as a series of wiggles—rather like a barometric graph of atmospheric pressure changes, or a seismograph recorder for earthquake shocks. Like those two examples too, the output graphs could vary enormously, from smooth, through jerky, to massive wiggles of manic dimensions depending on the form of the networks above which they were perched.

It may be helpful to take a look at the two types of pyramid shown in Figure 35. As can be seen, the number of nodes and their arrangement is exactly the same. What differs here is the number and the pattern of the interconnections between them. In the first example, the structure shows many interconnections, implying that each species in the food pyramid has few specific food requirements. The top predator here, for example,

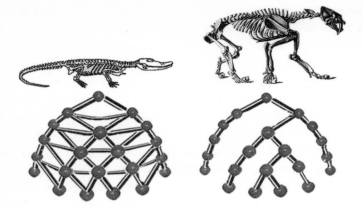

Figure 35. Two food pyramids are here constructed using a different pattern of network connections. Each has the same number of nodes (such as species) and the same general structure. In the left food pyramid, the top predator is envisaged as a generalized carnivore and there are many interconnections. In the right food pyramid, the top carnivore is much more specialized, as is the whole ecosystem structure. Such food pyramids may be especially vulnerable to collapse.

could be something like *Crocodylius*, the crocodile, eating hogs, frogs, and fish. The latter might then live on rather unfussy creatures like snails, shrimps, and midges. And those in turn might feed on a wide range of pondweeds and plankton.

In the second example, the number of available niches is much the same but the overall structure shows considerably fewer interconnections, implying that each node—each species in the pyramid—has much more specialized habits. In this case, our top predator might be envisaged as something a bit specialized, maybe like the extinct sabre-toothed cat *Smilodon*, a predator that we can fancifully imagine feeding fussily upon something rather specific—say, the elbows of a woolly Mammoth and the kneecaps of a Giant Sloth.[4]

Next, we looked at the output patterns—these being the signals about environmental change being received by the top predator. As we ran the models, we started to observe four different kinds of output patterns, shown in Plate 16. The first was smooth and flat; the second much noisier—a series of spikes; the third had the step-like appearance of a modern city skyline, while the fourth was very spiky, like a dense pine forest.

But which of these output patterns belonged to which pyramid? At first, it seemed hard to predict how each kind of pyramid would behave. But after running the models for a while, it began to emerge that our model ecosystems behaved in ways that reflected the patterns of interconnections between the nodes in each pyramid. Those pyramids with lots of interconnections tended to behave in a stable manner, with a smooth pattern of output (Plate 16, at left). Those with a minimal number of lateral connections tended to show an output that looked like a pine forest—lots of shocks reverberating through the system (Plate 16, at right). Those with an intermediate number of connections tended to show step-like behaviour.

At this point, purists may well quibble that no living ecosystem can behave exactly like the pyramids shown in these models. That is of course true. These pyramid models are far too simplistic. Models like these are little more than maps that help us to mimic reality. They are not part of the reality itself. But such a caution is true for all models, and most of all, I would argue it is true for the most sophisticated models.

What these models can help to show us is how to think. And most especially they show us how to think about the way that ecosystems can behave. They display patterns of behaviour that we could barely hope to guess. And those patterns lead us towards some startling conclusions about mass extinctions in the past, and in the future too. Food pyramids need not collapse because of drastic changes to the inputs alone—such as those that might be caused by ice ages, bolides, and volcanoes. They can be destroyed merely by the patterns of connection that exist within the pyramids themselves. Nothing more is needed. Some structures can be made unstable by just a push, like a house of cards.

The Albert and Emily effect

Computer models such as these clearly warn us about the hazards of simplistic thinking with regard to systems collapse, as during mass extinctions. Extrinsic causes are probably not enough to cause system collapse on their own, and they may not even be necessary.[5] That, alone, is a valuable insight. But it does not give us a feeling for the influence of geological time upon the build-up of ecosystem complexity and then its collapse. What we also require here is a geological sense of proportion. To that end, we will therefore need to look at yet another simile—the saga of Albert and Emily. The story which I have concocted goes something like this.

Once upon a time there lived a husband and wife in a block of flats. Albert was a biggish man, given to walking about in a string vest and a pair of red braces. Emily was a dainty little thing, who liked to read poetry while sitting on a *chaise longue* and sipping a glass of cream sherry. All had been going well enough between them until the fateful day that Albert took to smoking cigars in bed. Not little Wilhelm cheroots, but big Havana cigars that filled the room with fumes.

What could the poor lady do? At first, she took to making him stroll out on to the balcony to smoke his cigars. Then one day, she could stand it no longer, and she gave him a gentle push—and over the edge of the balcony went Albert. Down he fell with great force, to hit the flower bed beneath with a thud, stubbing out his Havana cigar and snapping off his braces.

Fortunately, the couple only lived on the first floor, and so Albert was able to clamber back up the stairs to the flat and dust himself down, while feeling more than a bit puzzled. A few weeks later, however, a very nice little flat became available on the tenth floor. So in they moved—'to enjoy the beautiful view from the balcony', as Emily so charmingly put it.

Then on one fateful evening, Emily repeated the same operation—the same gentle push from the same kind of balcony. Over the edge tumbled poor Albert, wearing the same kind of string vest and the very same kind of braces. And thud went the flower bed, putting out the same kind of cigar. But out went Albert too, stubbed out for ever.

So what killed Albert? This riddle can help to clear our brains of much nonsense about the collapse of ecosystems and their causes. It cannot have been the push from Emily, because exactly the same kind of push barely hurt Albert at all when they lived in the first floor flat. Some may scowl at this conundrum, because it conflicts with their notions of the way the world ought to work—it highlights a curious disconnection between cause and effect. Or more precisely, between an apparent cause and a real effect.

To be truthful, then, poor Albert did not die because of the magnitude of the push he was given. But because of the potential energy stored within his system on the tenth floor. He died because of the *state of the system* in which he lived. The higher he climbed, the harder he fell. And so it may be with ecosystems too.

Some kinds of system tend to have much higher levels of energy stored up within them than do others. Some organisms effectively live on the ground floor, while others live on the top floor. We might dare to say, for example, that Green Water ecosystems tend to live near the ground floor while Blue Water ecosystems come near to living life on the top floor. That might seem strange because many people still tend to think of ecosystems as all much the same, if they think about them at all. Plants capture sunbeams and make food; animals eat the plants, and then we eat animals. End of story. But it is much more complex and very much more interesting than that. For this is where we come face-to-face, yet again, with puzzles about the nature of the energy flow between two organisms, and within an ecosystem.

Efficiency is fatal

As in the world around us, there seem to be at least two distinct games being played here: the short game, and the long game. For example, some people find themselves in a world where they feel obliged to play for any short-term advantage—as with rock stars, footballers, and drug dealers. And, of course, bankers. Others find themselves in a world where they are obliged to hedge their bets in another way, and play the long game—as with planning a family. In the short term, raising a family makes almost no economic sense. It is only in the long term that its advantages can accrue. In the case of ecosystems, long periods of stability and hence of predictable conditions encourage ever greater efficiency and specialization—perhaps to the point where there is almost no slack left in the system. Every resource is shared and channelled and the system is exquisitely dependent on this finely tuned state. But when conditions change, there is nowhere left to run, meaning that the specialists become extinct: maximum efficiency proves fatal.

The killer here appears to be long periods of stability before a crash. In the fullness of time, there needs to be some slack still left in the system. If no slack is left, then collapse will take place. Looked at in this way, it may be unsafe to say that Darwinian natural selection of species is truly about 'survival of the fittest'—that is, if we take 'fittest' to mean 'the most efficient'.[6] For highly efficient things will always fail, in the fullness of time.

Seen in this light, we must come to the conclusion that mass extinctions are what big modern complex ecosystems do. Indeed, it is one of the things they do best of all. That is because efficiency is a major driving force for the development of big and complex systems. And greater efficiency encourages greater specialization. So, in the fullness of time, catastrophic collapse within any complex and highly structured system seems an almost inevitable outcome, including extinctions of both cellular organisms and human societies. System collapse is arguably, therefore, the price that must always be paid for complexity.

In-breeding

Does the fossil record also provide us with lessons about the fate of our own increasingly complex and interlinked society? I think it does. Consider, for example, the dangers presented by a growing dependence upon information networks and energy supplies. Or the computer games we have just been playing. A hundred years ago, there were few such partnerships, meaning that a failure in the early power supply would have had scant effect on the world. Most of society ran on local coal or wood supplies, and only a district or two would suffer from a local shortage. The rest of the world could therefore sail on as before. But our present societies are much more closely interlinked. Even the fuel supplies for electric power stations travel vast distances between nations.

Imagine, then, the influence of a highly stable and predictable fuel and electricity supply that might last for hundreds of years in a row. Economies, like symbioses and ecosystems, could become finely-tuned, by 'Darwinian' market forces, towards the very edges of efficiency. They would be driven by efficiency-savings to climb upwards 'from the first floor towards the tenth floor'. It is now easy to envisage what might happen to this highly interconnected world when the power supply eventually fails, as probability predicts. Or when a computer virus runs rampant, as chance dictates. Complex systems might then collapse. As with the fossil record, social partnerships would be dissolved by a trauma to the energy supply.

Further possible parallels with the fossil record are intriguingly common. For example, collapses in financial markets typically take place after long bouts of stability. Ultimately, the latter prove disastrous for both speculators and debtors. But there are other fields where society may need to stay on its guard as well. The list here includes: a loss of diversity in the sea, including reefs; a loss of diversity on land, including rainforests; a loss of diversity in healthcare; too much reliance on limited sources of water, or on limited sources of oil and gas; in-breeding programmes for crops and livestock; and in-breeding within institutions and political leaders. Each of these involves a loss of diversity between connections.[7] And that means that the killer, when it comes, isn't so much 'the push'. The real killer is the complacency: the lull that comes before the storm, as with any bout of misplaced confidence.[8]

The push by Emily from the balcony need not prove fatal, though. If some elasticity is left in the system, then Albert may survive. In other words, 'inefficiencies' are not always a bad thing within a complex system. In the longer term they may even be a good thing, like unharvested patches of coral reef left alone on the sea floor. Or family silver left alone in the bank; or assets left unstripped in the public domain. Lots of little inefficiencies—tweaked by many little mishaps—may help to moderate the effects of mayhem, when it comes.

The crunch

The regularity of mass extinctions in reef ecosystems seemingly reflects the development of specialized interconnections between its symbiotic associations. The fossil record suggests that 5 to 10 million years may not have been quite enough time to tie bonds together to the point where stresses can bring about complete extinction of both host and symbiont. But 20 to 30 million years without perturbations seems to have been long enough to lead to some wholly dependent partnerships—and then to their doom. As with poor Albert falling down from the tenth floor, the whole ecosystem then becomes liable to collapse. That in good measure explains why we no longer find the fossils from the Sphinx still living on the seafloor. Those were old liaisons, forged long ago, and long ago dissolved by 'events'.

Now we are getting to the core of the question. So far as we know, no new organelles have arisen from symbioses between cells during the last 540 million years or so. A major reason for this lack of novelty may be that 20 to 30 million years is simply not enough time for genetic transfers to take place between guest and host before the arrival of the next trauma. Being the owners of a workable genetic code, the symbionts can still flee to safety.

Both living and fossil symbioses reveal that the path towards the loss of freedom by captured organelles may have been a long and difficult one. The evolution of the modern eukaryote cell—by means of symbiosis—may actually have required much longer than 30 million years for the entrapment of these structures. Maybe, indeed, it required *huge* amounts of time—say, 100 to 1000 million years—during a time of considerable environmental stability. Only then could guest and host have sunk their differences to develop into a wholly new kind of organism. The Boring Billion neatly fits the bill.

Planets X77 and Y12

As we scan the Galaxy in search of other habitable planets, we might pause to wonder whether such a story is unique to planet Earth. Could it be that there are other planets out there that gave birth to life which otherwise had quite different trajectories? Consider for example, the Planets X77 and Y12.

X77 is a hypothetical planet much like Earth, but its surface is a more restless place. Bacteria abound there. But its simple cells have never reached the stage of eukaryote cells—with nucleus, mitochondrion, and chloroplast—because these associations keep getting wiped out by massive asteroid collisions. No plants can be found on Planet X77, alas.

Y12 is a very different planet. Somewhat like Earth, its surface is an altogether calmer place. No big asteroid collisions, and no great glaciations. Bacteria therefore abound, while many of its cells have now reached the level of eukaryote organization, replete with cell organelles. But a closer look shows that things have gone a bit out of control here. Much of the life on Planet Y12 seems to be made from giant cells in which almost all their needs are being met internally. The surface of the planet is covered with a pulsating mass of interconnected slime which can slither up into the sunshine or down into the rocks according to local need. There are no mass extinctions here. Nor are there animals or plants. Life has approached a final state of one-ness. A single being free from accidents.[9]

Neither Planet X77 nor Planet Y12 have yet been found of course. But they might be out there somewhere.

Bladderwrack

Gairloch on a late autumnal afternoon is a stirring place. I had travelled to this coastal village in the far northwest corner of Scotland to collect some rocks, and to collect my thoughts. In the cliffs at my back were crushed remains of banded iron formations, much like those we met in the Gunflint chert of Canada with their bacterial cells almost 2 billion years old. Along the shore to the north were the remains of much younger beds—Torridonian sediments around a billion years old (Plate 15), and in these we were starting to find much more in the fossils than simple bacterial cells. We had come across complex and wondrously preserved eukaryote cells, much like those in the 1.4 billion year old Roper Group of Australia. But these putative Green algae had not lived and died in the sea. They had bloomed in the waters of lakes on land that existed a billion years ago: the Torridonian rocks contain some of the most ancient lake beds ever found. Our results were showing how complex life had started to green the surface of continents at least a billion years ago.[10]

The scene along the shore was suitably atmospheric, too. White tails of spume were spinning off the tips of the surf, and breakers lashed the shore to form a mash of froth, foam, and brown bladderwrack. Out to sea, and beyond this mash of kelp, I could see two great battleships ploughing through the waves beneath an azure sky streaked with white cloud. The ships were a reminder that it was from hereabouts that sailors set out on Arctic Convoys towards Russia during the later stages of World War II. And it was hereabouts, too, that some of the first ever experiments in germ warfare had been conducted. On Gruinard Island, sheep had been injected with *Bacillus anthracis*—anthrax. Monkeys had even been launched on a raft nearby, replete with infections of *Yersinia pestis*—the bubonic plague bacteria with which our story began.

Scrambling down to the nearby shore, the sea's hiss turned into a full-throated rumble. Far off, the battleships were now rolling sickeningly through the spume. I could well imagine the scene on board: the captain standing sternly amid his chartroom; petty officers anxiously guarding their propellers; smashed crockery rolling across the sick bay floor; a world filled with secret chambers. Life on board ship may indeed resemble the life of a cell, but I was glad to be ashore.

Down on the beach, by my feet, the mash of bladderwrack was building higher and higher, brown and glistening. Peering closely, I could spy the tiny white coils of seashells, still holding on for dear life—the secret chambers of Foraminifera, just like those first seen by Robert Hooke back in 1665. Here were some distant relatives of the fossils first reported by Strabo beside the Pyramids and the Sphinx as early as 20 BC. These were clues to the riddle of the Sphinx Within; of symbiotic worlds turned outside-in. And then turned inside-out again during mass extinctions.

Here and there within the mound of kelp were strange-looking fronds, recalling those fright wigs of *Sargassum* I had seen floating on the Sargasso Sea, but now washed ashore by a storm. Each frond was built from thousands of cells. And each cell contained an organelle—the chloroplast—whose ancestry dated back to the Boring Billion, between 1 and 2 thousand million years ago; a time when the world lacked booms and busts, a time when the Earth seemed eerily calm. Without that exotic interval, this greening of our planet might never have happened. And we would not be here to ponder it.

NOTES

PREFACE

1. For discussions on this theme of the Cambrian explosion, see Brasier 2009 and references therein.

CHAPTER 1

1. Some 69,000 died during this last great epidemic of bubonic plague in England. The ability of church parishes to deal with this mass of dead rapidly broke down into chaos, and many bodies were cremated rather than buried.

2. Lofty ideals for the pursuit of science were put forward by Sir Francis Bacon in his *New Atlantis*, published in 1627 after his death. In this article, Bacon conjured up the prospect of a college dedicated to science, which he quaintly called Saloman's House, meaning the house of the fabled Jewish philosopher-magician, today called King Solomon. See Johnston 1965.

3. Sadly, this building no longer exists. Published engravings show it to have been set out like an Oxford college such as my own at St Edmund Hall, or like one of the Inns of Court near the Temple Church in London. Those, in turn, followed the plans of Carthusian monasteries, and Roman villas.

4. The fossil collection of the Royal Society was originally stored in the rooms of Robert Hooke at Gresham College. See Jardine 2004, p. 39.

5. *Micrographia* by Robert Hooke was first published by the Royal Society during the Plague year, in January 1665. See Hooke 1665 for full details.

6. The Preface to *Micrographia* (Hooke 1665) carefully describes and illustrates such a lens-making machine.

7. See Hooke 1665, in his Preface to *Micrographia*.

8. Hooke was scarred by smallpox as a child. But the diarist Samuel Pepys wrote in February 1665: 'Mr Hooke...is the most, and promises the least of any man in the

world that ever I saw. Here was excellent discourse till ten at night and then home.' See Bright 1960a, Pepys in his Diaries II, 15 February 1665.

9. As he us tells is himself: 'I here presume to bring in that which is more *proportionable* to the *Smalness* of my Abilities, and to offer some of the least of all *visible things* to that *Mighty King*, that has *established an Empire* over the best of all *Invisible things* of this World, the *Minds* of Men.' See Robert Hooke 1665 in his Preface to *Micrographia*.

10. '...the Arts of life have for too long been imprison'd in the dark shops of Mechanicks themselves & there hindered from growth, either by ignorance, or self-interest.' See Hooke 1665, in the Preface to *Micrographia*.

11. Hooke writes in his Preface to *Micrographia*: 'towards the prosecution of this method in Physical Inquiries, I have here and there gleaned up a handful of observations, in the collection of most of which I made use of Microscopes, and some other glasses and Instruments that improve the sense; ...to promote the use of Mechanical helps for the Senses, both in the surveying the already visible World, and for the discovery of many others hitherto unknown, and to make us, with the great Conqueror, to be affected that we have not yet overcome one World when there are so many others to be discovered, every considerable improvement of Telescopes and Microscopes producing new Worlds and Terra-Incognita's to our view.' Hooke 1665.

12. Hooke was not the first to make use of a microscope. The earliest compound microscope was reputedly made by Sacharias Jensen, a spectacle manufacturer and counterfeit coin-maker from Middleburg in Holland, around the year 1595. See Stewart 2003.

13. Hooke 1665, Preface to *Micrographia*.

14. See Hooke 1665, conclusion to the Preface of *Micrographia*.

15. Hooke tells us 'That in divers of them the microscopic objects the Gravers have pretty well follow'd my directions and draughts; and that in making them, I endeavoured (as far as I was able) first to discover the true appearances, and next to make a plain representation of it.' See Hooke 1665, in the Preface to *Micrographia*.

16. On 2 January 1665, before the arrival of the Plague, Diarist Samuel Pepys wrote: 'Thence to my bookseller's, and at his binder's saw Hooke's book of the Microscope which is so pretty that I presently bespoke ordered it, and away home, where, thinking to be merry, was vexed by my wife's having looked out a letter in Sir Philip Sidney about jealousy, for me to read, which she industriously and maliciously caused me to do...and then to bed.' See Pepys in his Diaries II, 2 January 1665 (Bright 1960b).

17. Hooke 1665, *Micrographia*, Observ. XXI. Of Moss, and several other small-vegetative Substances.

18. The idea of spontaneous generation had been around for millennia, but Hooke was among the first to speculate about it scientifically, writing 'that Mould and Mushroms require no seminal property, but the former may be produc'd at any time from any kind of *putrifying* Animal, or Vegetable Substance'. Hooke 1665, *Micrographia*, Observ. XX. *Of* blue Mould, *and of the first Principles of Vegetation arising from* Putrefaction.

19. See Hooke 1665, *Micrographia*, Observ. XX. *Of* blue Mould, *and of the first Principles of Vegetation arising from* Putrefaction.

20. Experiments by Italian physician Francesco Redi in 1668 were among the first to refute spontaneous generation. Even Schwann and Schleiden in 1838 invoked something close to spontaneous generation when trying to explain how new cells arose.This old idea was finally demolished by Louis Pasteur in 1860, who was able to claim a prize of 2500 Francs from the Academie de Sciences Francaise.

21. Hooke's microscope was much like an early telescope, having an eyecup to maintain the correct distance between the eye and eyepiece, separate draw tubes for focusing, and a ball and socket joint so as to incline the body. For the optics, Hooke used a bi-convex objective lens at the bottom, combined with an eyepiece lens at the top, and a tube- or field-lens in the middle. Such a combination had an unfortunate effect, causing significant chromatic and spherical aberration of the images seen down the tube, which Hooke attempted to correct by means of diaphragms and other devices. American Physical Society website. See http://www.aps.org/publications/apsnews/200403/history.cfm.

22. Hooke 1665, Observ. LIII. *Of a* Flea.

23. See Hooke 1665, Observ. XVIII. *Of the* Schematisme *or* Texture *of* Cork, *and of the Cells and Pores of some other such frothy Bodies.*

24. Hooke 1665, Observ. XVI. *Of* Charcoal, *or burnt* Vegetables.

25. Sadly, Hooke became distracted, and his findings on fossils were not to be published until after his death. See Jardine 2004.

26. Hooke 1665, Observ. XVII. *Of* Petrify'd wood, *and other* Petrify'd bodies.

27. See Hooke, *Micrographia*, Observation XI. *Of* Figures *observ'd in small Sand.*

28. For a simple introduction to fossil Foraminifera and other protozoans we shall meet, see Brasier 1980, and Armstrong and Brasier 2004.

29. Quoted from Diary of Samuel Pepys, 25 June 1667 (Bright 1960b). Oldenburg was the first Secretary to the Royal Society and was therefore based at Gresham College alongside Hooke. It seems that Hooke was also much envied by another young scientist in France, called Adrien Auzout, who complained that Hooke's lens-making engine was not adequately demonstrated in the pages of *Micrographia*. Hooke

had also upset a powerful adversary in Holland—Christiaan Huygens—who was one of the grand masters of optical theory. Both Auzout and Huygens were, it seems, in receipt of letters of intelligence from Sir Robert Moray and Henry Oldenburg concerning Hooke's experiments. See Lisa Jardine 2009. See also Robert Lomas, 2007, Gresham College website, http://www.gresham.ac.uk/professors-and-speakers/dr-robert-lomas, last accessed 11 January 2012.

30. Newton was familiar with his inner beast. He penned Masonic musings, including one called 'The Power of the Eleventh Horn of Daniel's fourth Beast, to change times and laws'. See White 1998.

31. Newton is even reputed to have removed pictures of Hooke from the Royal Society. See Jardine 2004.

32. As Emily Dickinson once put it, back in the mid-nineteenth century: 'Faith is a fine invention, When Gentlemen can see—But Microscopes are prudent, In an Emergency.' See McNeil 1997.

33. See Jardine 2009. See also Robert Lomas, 2007, Gresham College website, note 29.

34. Antonie van Leeuwenhoek—often mistakenly called the inventor of the micro-scope—was actually looking for ways to detect faults in linen fabric. It was developed from the simple blob-of-glass lens developed and described by Hooke in 1665 (see Hooke, Preface to *Micrographia*). Van Leeuwenhoek's microscope could magnify as much as 270 times. With such a powerful tool, van Leeuwenhoek was able to deliver a plethora of publications in the years up to 1723. He was the first to describe forms that we would now call algae and protozoans back in 1674, followed by bacterial forms in 1683. He even observed blood cells and reported on the first spermatozoa—reputedly his own. Observations such as these were written in Dutch and then trans-lated into English for publication in the *Philosophical Transactions of the Royal Society*.

35. By a happy chance, my first ever microscope was of this kind. I have it still. Appropriately, this Christmas present circa 1958 came provided with a flea mounted on a glass slide.

36. Joseph Jackson Lister of London was reputedly the first to help make the compound microscope into a powerful biological tool, publishing his results with the Royal Society in 1830. See Lister 1830.

37. At this time, and before the 1950s in fact, the term cells always meant what we would now called eukaryote cells—nucleated cells. Bacterial cells took other names, such as cytode or moneran. See Fet and Margulis 2010.

38. For a more detailed account, see Vasil 2008.

39. Schleiden and Schwann found that all living things are made from one or more cells; and that the cell is the basic unit of structure for all organisms. By 1855, Rudolf

Virchow was able to complete the three modern statements of Cell Theory, when he found that all cells come from pre-existing cells, famously encapsulated by him in the Latin phrase *omnis cellula e cellula*. See Vasil 2008, Hardin et al. 2012.

40. For early concerns about the nature of individuality in cells, see Huxley 1912.

CHAPTER 2

1. For an excellent account of this period in Darwin's Life, see Browne 1995.

2. Euston Road, just to the north, was reputedly an area dominated by clay pits, brick-works, and huge rubbish tips where cinders and rubble was sorted out and sold to the brickmakers.

3. Also present were John Henslow, William Fitton and their wives, see Browne 1995, p. 412.

4. Although his book called *On the Origin of Species* would not be published for another twenty years, in 1859, Darwin was already making notes on his theory of evolution. See Darwin 1859 and Browne 1995.

5. The full title was: *Journal of researches into the geology and natural history of the various countries visited by HMS Beagle*. The book was submitted in 1838 and published in 1839. Darwin had also been making some jottings about the origin of species at about this time, but he kept these largely to himself. He was clearly concerned about their appearing too speculative, like the writings of his grandfather, Erasmus Darwin, who was a noted atheist and something of a sexual libertarian too. See Browne 1995.

6. See Darwin, 1842. The manuscript for this book was ready by 1840 but its publication was delayed for two years by ill health. This was his first conceptual book.

7. Darwin's major contribution was his use of coral reefs for testing an hypothesis about the rate and nature of changes to the Earth and its biosphere on geological time scales. See Darwin 1842. Lots of earlier observations had been made, of course, such as by the great naval officer and Australian explorer Matthew Flinders who studied the Great Barrier Reef from his ship *Investigator* in 1802. On his way back to England in 1803, Flinders was captured by the French and imprisoned for more than six years. He didn't arrive home until 1810 and was by then a very sick man. Lamentably, he died the day after his book *A Voyage to Terra Australis* was published in 1814. See Flannery 2001.

8. This was one of Darwin's many important observations. Until his research, it had been thought that reef-building corals could live at all manner of depths. That is only true for non-reef building corals, for reasons we shall meet later. See Darwin 1842.

9. Charts available to Darwin include those made by ships from Britain, France, the East India Company, the Netherlands, and the USA.

10. This angle often exceeds 45 degrees, and as Darwin himself realized, this angle was far too steep to be accounted for by the build up of loose sand from the bottom. See Darwin 1842, p. 23.

11. Darwin 1842, p. 147.

12. See Herbert 2005.

13. That geology was the queen of sciences in the 1830s and 1840s has been shown by science historians Pietro Corsi and Simon Knell. See, for example, Knell 2009.

14. Some might object to this thought because the co-author of the Darwinian theory—Alfred Russel Wallace—was never a geologist like Darwin. But Russel Wallace was a founding father of biogeography which, like geology, is forced to grapple with the distribution of species in both space and time. For more on Russel Wallace and biogeography, see Smith and Beccaloni 2010.

15. Heretical because the hand of God was usually seen, back in the nineteenth century, as the prime first cause. It was the Cambridge philosopher William Paley who firmly encapsulated that view. Other Cambridge dons, such as Adam Sedgwick and William Whewell, clung to a similar view as well. But we can see that Darwin was already moving away from Paley's Blind Watchmaker towards the arguments in the *Origin of Species* in 1842: 'In an old-standing reef, the corals, which are so different in kind on different parts of it, are probably all adapted to the stations they occupy, and hold their places, like other organic beings, by a struggle one with another, and with external nature; hence we may infer that their growth would generally be slow, except under peculiarly favourable circumstances.' See Darwin 1842, p. 76.

16. Interestingly, Darwin did not come up with his theory on coral reefs after visiting a coral reef. He came up with it beforehand, whilst climbing aloft in the mountains of the Andean mountains of South America. He realized that all the evidence pointed to these islands 'evolving' long before he had arrived on the scene. And that they would continue going through their motions long after he had gone—which indeed they have. By looking and thinking in this way during the cruise over five long years, Darwin had stumbled on a fundamental truth. A fact that was to prove essential for our narrative: that the birth and death of reefs carries a remarkable record of evolution—both planetary and ecological—that extends very far back in time. For more on this, see Darwin 1842 and Browne 1995.

17. Darwin over-emphasized geological subsidence in his book of 1842 because the evidence for an alternative cause of flooding—post-glacial sea level rise—was essentially unknown. Although opposed by Alexander Agassiz (see David Dobbs 2005), Darwin's work on coral reefs strikes many readers as remarkably accurate

and perceptive, even today. After many abortive attempts, Darwin's theory of coral atolls was at last confirmed in 1952, whilst drilling at Bikini for the atomic bomb tests, as we shall later see.

18. From 1801 to 1805, at the instigation of Sir Joseph Banks, he collected over 2000 new species of plant from Australia, working closely with captain Matthew Flinders. Unhappily, many specimens were lost when *HMS Porpoise* sank on the return voyage. In 1810 he published his famous *Prodromus Florae Novae Hollandiae et Insulae Van Diemen*. See Mabberley 1985.

19. Brown was Keeper of Botany at the British Museum from 1837 until his death in 1858.

20. It has been claimed by some that Brown was unable to see his own Brownian motion using his simple microscope. This has since been convincingly rebutted by Brian Ford (see Ford, 1992). Robert Brown's own microscope, made famous by his studies of cells and Brownian motion, can still be seen on display in the rooms of the Linnean Society at Burlington House, in London.

21. His paper was read before the Linnean Society in 1831 and published in 1833. See Mabberley 1985.

22. For some reason, nuclei are especially clear in this group of flowering plants, the monocotyledons.

23. Charles Lyell 1850, p. 565 cited Robert Brown thus: 'the structure of the husks or that part of the flower which is persistent, agrees precisely with the barley of the present day . . .'. Modern work can find differences in the strain of cultivated plants but not in the species, of course, since the time elapsed is so short, a mere 4000 years.

24. My own italics. He reached this view in 1868. See Huxley 1898, p. 138–42.

25. See Brasier 2009, pp. 223–5.

26. See Haeckel 1879. The term 'Monera' was used in various ways by Haeckel, to include not only bacteria, but also protozoans and even sponges. It took another century for the term Monera to refer to the bacteria alone (See Whittaker 1969, 150–60; see also Margulis 1998, p. 61). We should also note that Haeckel's usual terms for the non-nucleated cells of bacteria and cyanobacteria were actually 'cytode' and 'bioblast'. See Fet and Margulis 2010.

27. A contemporary overview was given by Allman in his Presidential Address of 1879 to the British Association for the Advancement of Science of 1879. See Allman 1879.

28. Plaudits for this refutation are conventionally given to the French microbiologist Louis Pasteur. See also Note 20, p. 221.

29. By 1855, for example, we find this within a student textbook by Balfour (1855, p. 5): 'The cell may be considered the ultimate structural element of all organisms'. These early revelations were to lead towards a remarkable succession of findings, including the first statements of cell theory by Verchow in 1858, the discovery of DNA by Miescher in 1869; the first observation of chromosomes by Alexander Flemming in 1877; the distinction between non-nucelated prokaryotes and nucleated eukaryote cells by Edouard Chatton in 1925; decoding of the double helix by Francis Crick and James Watson in 1953; and most recently, by the molecular evidence that all of life— including humans and apes—are cousins within a single Great Tree of Life. See Hardin et al. 2012; Fet and Margulis 2010.

30. We know this from the correspondence of Emma Darwin. See Litchfield 1904, p. 461.

31. So filled with wisdom and common sense is this great book, that were every modern treatise on geology to be burned by zealots, his writings would arguably be enough to salvage the science—and humanity with it. It is filled with modern obsessions too, like observations on man-made climate change and extinction. See Lyell 1837 and Lyell 1850.

32. Lyell devoted much of the second (1832) volume of his *Principles of Geology* to this problem. He even devoted over 20 pages to a discussion on the origin of coral reefs in his fifth edition of the *Principles of Geology*. Lyell 1850.

33. With the advantage of hindsight. The island of Santorini was a volcanic rim with a central crater lake flooded by the sea. The Minoan civilization of ancient Crete used Santorini as a major trading port. A stupendous eruption around 1600 BC (carbon 14 date), along with its attendant tidal wave effectively destroyed harbours and brought much of Minoan power in the Mediterranean to an end. See Bottema 2003, Freiderich et al. 2006.

34. Lyell himself made oblique reference to this: 'I am not about to advocate the doctrine of general catastrophes recurring at certain intervals, as in the ancient Oriental cosmogonies.' Lyell 1850, p. 668.

35. The 'indifferent god' argument was gaining ground in Europe after the catastrophe caused by the Lisbon earthquake in 1755. See, for example, Araujo 2006.

36. And if he wasn't an omnipotent force, then why call him God? That was, and still is, the view of Deist scientists, for whom God was simply the founding force behind the rules of our universe. This conundrum of three conflicting statements—Good is Good— God is all-powerful—But Bad things happen—is known as the Theodicy problem.

37. Pietro Corsi has made the case that the British establishment was worried that the French Revolution of 1830 might ignite a nation already inflamed by the lack of Reform. There was, remember, a further series of revolts in 1848. See Corsi 2005.

38. I have here paraphrased Lyell 1850, chapters XLII to XLIV.

39. In this eternalism he was not only following the writings of James Hutton (e.g. Craig 1997; McIntyre and McKirdy 1997) but arguably also the philosophy of Aristotle of ancient Greece.

40. Lyell papered over some damaging inconsistencies in his thinking. Prof. Pietro Corsi of Oxford has found that William Whewell of Cambridge had been taunting Lyell in 1836 with the following paradox: if there are mass extinctions in the fossil record, from whence come their replacement species? If they come from God, then the Earth cannot be visualized as a godless system of the kind assumed in Lyell's Principles, since the Creator clearly intervenes on the planet from time to time. This seems to be what the geologists Sedgwick, Murchison, and Phillips tended to believe. But if these new species did not come from God, then they must have emerged from within nature itself by a process of evolution. This is, of course, what the Frenchman Lamarck had been arguing since 1809. But Lamarck's godless universe was a hypothesis that was fiercely contested by Lyell through some 200 pages of the *Principles*. This conundrum meant that Lyell found it hard to gain powerful followers, with one curious exception—his young protégé, Charles Darwin.

41. A few hundred metres south of Darwin's house stood the slums of 'the Rookery' where street battles raged, even in 1840. One in eight of London's criminals lived there. See Palm-Gold 2011.

42. See Litchfield 1904, p. 461, and Browne 2003, p. 412.

43. This curious rock can now been seen on display at Darwin's house at Down in Kent.

44. Darwin's book on Coral Reefs in 1842, neatly exemplified his Lyell-like approach, from studying patterns (in chapters 1–4, plus a very extensive Appendix), towards a teasing out of the process with the greatest explanatory power (in his chapters 5 and 6). Like Charles Lyell, he also chose to work backwards through the time, like an archaeologist digging through layers of dirt (see Lyell 1832, 1837). It is this same structure of argument that Darwin later used in 1859 in his *Origin of Species*. See Darwin 1842 and 1859.

45. My italics. Darwin was writing here about the distinctive relationship he was finding between living and extinct creatures on different continents such as South America and Australia. See Darwin, 1845. Interestingly, this quote does not seem to be present in the original, 1839, edition.

CHAPTER 3

1. This is reminiscent of a quote from the famous poem of 1798 by Samuel Taylor Coleridge, 'The Rime of the Ancient Mariner': 'Day after day, day after day,/

We stuck, nor breath nor motion;/As idle as a painted ship/Upon a painted ocean'. The poem may have been inspired by James Cook's second voyage of exploration (1772–1775) aboard *HMS Resolution*; Coleridge's tutor, William Wales, was reputedly the astronomer on Cook's flagship. See Keach 1997.

2. Born 1748, died 1832. See McKnown 2003 for a collection of Bentham's famously atheistic views.

3. The Auto-Icon first appeared in the Anatomical Museum of UCL in 1850, moving to its present place in 1965. The body was at first dissected in public and the bones were then wired together, while his mummified head used to be displayed beneath his feet. It was later stored separately, to discourage its use as a trophy by rival gangs of students from different London colleges. Such mascots caused much disruption and were progressively hidden away by the authorities. In 1968, I was actually part of a team that unearthed a long lost mascot of King's College—a public house Red Lion with bright blue testicles, affectionately known as Reggie. It had been dumped in the woods near Dorking in Surrey. Sadly, we Chelsea Boys were debarred from marching this trophy through the streets of London.

4. It was also the place where I first lectured on aspects of the two chapters that follow, to the Annual General Meeting of The Micropalaeontological Society, in 1981.

5. Sir W.M.F. Flinders-Petrie (1853–1942) was a famous Egyptologist who, as his name implies, was a direct descendent of the Australian explorer Matthew Flinders. He is widely regarded as the founder father of modern archaeology. See Drower 1995.

6. Much of my research life seems to have been spent in huts and hovels. After UCL, I moved to the British Geological Survey in Leeds, where we worked in an old wartime hospital of asbestos sheds. From there I moved to Reading, to work in a prefabricated outbuilding. And from thence I moved to Hull, where I was placed in a wooden workman's hut for two years. My start at Oxford was hardly much better—two years spent at the top of a corkscrew staircase in a garret once used by the Natural History Museum printing press. And then two years in temporary quarters beside the *Dodo*. But I was so absorbed in my studies, that little of this ever seemed to matter much.

7. I was 22 years of age when I joined my ship, coincidentally like Darwin in 1831, Joseph Hooker in 1839, and Thomas Huxley in 1847. Darwin was on *HMS Beagle*; Hooker on *HMS Erebus*; and Hooker on *HMS Rattlesnake*. For stories from these older and more august sailors, see McCalman 2010.

8. *HMS Fawn* was built at Brooke Marine in Lowestoft and launched in October 1968. She had a displacement of 1160 long tonnes, a length of 58m, a beam of 13m, a speed of up to 15 knots and a complement of 44 men. Sadly, in 1991 she was decommissioned and sold for commercial surveying.

9. Another colleague, Peter Wigley joined us later, but he had fewer dealings with the navy.

10. Apollo 12 was launched on 14 November and the returned on 24 November 1969.

11. I was to share quarters with one of the Navigating Officers, Lieutenant Simon Richardson. Other officers on the ship included Lou Davidson and Matthew French.

12. My mother, father, and grandmother had travelled down from Torquay. I wasn't to see them or speak to them again for the rest of the year. That's how it was back then.

13. For an introduction to the science of some of the seaweeds we shall meet, see Thomas 2002.

14. This is now regarded by some as a variety of the Green alga called *Ulva*, but I have followed Thomas 2002 in using the older more evocative name.

15. For example, huge blooms now afflict the China Sea and the Baltic Sea.

16. This tubular shape has a lower surface area to volume than a thin sheet, and this helps it to reduce water loss at low tide. See Thomas 2002.

17. Chloroplasts are sometimes called 'plastids', in part because they are not always green in colour, but I will stick to this more widely used term. Their green colour is due to pigments called chlorophylls *a* and *b*, plus smaller amounts of reddish carotene and brown xanthophylls. Phycobilin pigments are absent.

18. The Green algae are usually placed in the Division Chlorophyta.

19. The green pigments of chlorophyll absorb light between the wavelengths of 400–500nm. See Thomas 2002.

20. The Sargasso Sea was named by a Portuguese sailor from the likeness of its seaweed to a species of *Helianthemum* that thrived in the wells and waterholes of his homeland, where it was called 'Sargaco'.

21. Photosynthetic pigment of Brown algae include chlorophylls *a* and *c*, together with carotenoids and xanthophylls, including fucoxanthin, a brown-coloured pigment that gives them the dark brown colour.

22. The Brown algae or Phaeophyta, comprise about 2000 living species. They are currently placed together with diatoms and water moulds within the Heterokonta, or Stramenopiles. See Hoek et al. 1995.

23. The brown pigments of β-carotene and fucoxanthin can absorb light within the green part of the spectrum, 400–520nm. See Thomas 2002.

24. The ancestors of Brown algae may well have included members of the Yellow Green algae or Xanthophyta, a group mainly found in freshwater. So one possibility is

that the ancestors of Brown algae may have moved from the land to the sea, perhaps in the last 250 million years or so. See Raven et al. 2002.

25. Nelson's Dockyard lies within the inlet called English Harbour.

26. Other Blue Greens likely included in *Lingbya* and *Schizothrix*.

27. The genetic information within bacterial cells—of both eubacteria and archaebacterial kind—is commonly folded into a compact structure known as the 'nucleoid'. See Hardin et al. 2012.

28. There has been some move away from using the term prokaryote in recent years, in part because it is a negative feature defined on what these cells lack, and partly because it implies fundamental similarity between all organisms that lack a nucleus, which is questionable. See Hardin et al. 2012.

29. A distinction is commonly made between single celled eukaryotes, which are called Protista, and the multicelled groups which are called Fungi, Plantae, and Animalia. Single-celled protists include Foraminifera, as well as *Amoeba* and *Paramecium*, together with chloroplast-containing single-celled algae such as the diatoms, dinoflagellates, and *Phaeocystis*. In some classifications, both single- and multicelled algae and single-celled protists are lumped together into the Protoctista. In others, this bigger grouping is also called the Protista, which is what I shall call them here. See Margulis 1981.

30. The first observations on cell walls were of course those made by Robert Hooke on his cork tree cells in 1665. By the time that bark cells of *Quercus* had turned into cork, their contents had long dried away, and Hooke was mainly seeing cell walls, not whole cells.

31. The biological term for such regulation is 'homeostasis'.

32. Should you be in need of this information, gonorrhoea is brought about by *Neisseria gonorhoeae*. Archaeologists working on the Tudor warship *Mary Rose* suspect that naval surgeons have been using syringes against urethritic infections since the early 1500s.

33. Phospholipids are one of three kinds of lipid found in nature, the others being fats and steroids. See Novikoff and Holzman 1976.

34. Normal fats have a backbone formed from three glycerols. In phospholipids, one of these glycerol molecules is replaced by a phosphate-containing group. The cell membrane typically consists of two such layers of phospholipids. See Novikoff and Holzman 1976, Hardin et al. 2012.

35. These pores are called ion pores. They selectively allow ions in or out of the cell if they are the right size and have the right electric charge. Pore-like structures called membrane pumps, or transporters, are formed from special types of protein that use energy from ATP to move molecules and ions around. Carrier proteins will also bind to a particular solute and lead it into the cell. Active transporters can move materials against the trend, with the aid of energy.

36. This concept was reputedly first suggested by the Russian biologist, Vladimir Vernadsky.

37. This process of equalization between solute concentrations across a selectively permeable membrane is called osmosis. It is fundamental for life. It was first reported in 1748 by Jean-Antoine Nollet.

38. The others which seem to be essential for nearly all kinds of life include nitrogen, phosphorus, and sulphur as well as Na, Mg, Cl, K, Ca, Mn, Fe, Co Ni, Cu, Zn, Mo, and W. See Williams and Frausto da Silva 1996.

39. See T.C. Cavalier Smith 2006.

40. 'Gram positive' is a primary characteristic of mainly aerobic bacteria having a cell membrane composed of a thick layer of peptidoglycan with techoic acid. They retain the purple stain, even after decolourising. Cavalier-Smith (2006) regards this single-layered wall as a later feature of cell evolution. Because of the nature of this membrane, they stay purple with the stain. The bulk of Gram positive bacteria are found within the Actinobacteria and Firmicutes, a vast and diverse group of tiny rod-shaped cells, including the microbes that give to soil its distinctive smell.

41. Gram negative is a characteristic of bacteria having a cell membrane composed of a thin layer of peptidoglycan together with an outer membrane of lipopolysaccharide. Cavalier-Smith (2006) sees this two-layered structure as an early feature of cell evolution, but this is far from resolved. This structure causes them to lose the crystal violet—iodine stain on decolourising and take up the red counter stain. Further examples include spirochaetes, and the bringers of leprosy—*Mycobacterium leprae*—as well as tuberculosis—*M. tuberculae*.

42. Because they have a distinct history in their 16S RNA, Carl Woese suggested they are a sister group of the Eubacteria, which he called later the Bacteria (Woese and Fox 1977). He also called them Archaea (meaning old) because he thought they evolved before Bacteria. In contrast, Cavalier-Smith (2006) continues to call them Archaebacteria, and suggests that these evolved from a Posibacterial branch of the Eubacteria (such as the Actinomycetes) by means of further simplification of the cell membrane. In this view, the ancestral eubacteria had a murein-rich wall. The descendent archaebacteria then lost the murein in their cell wall.

43. Animal cells can also be reinforced by a strong but elastic network of collagen fibres. Hardin et al. 2012.

44. Phagocytosis means the eating of other cells. There is another but related phenomenon called pinocytosis, in which macromolecules are engulfed. Phagocytosis and pinocytosis together are called endocytosis.

45. As we have seen, it was Robert Brown who named the nucleus in plants; and Schwann and Schleidden who found it to be present in both plants and animals.

46. See Novikoff and Holzman 1976.

47. See Hardin et al. 2012.

48. The nuclear envelope has a double-membrane structure that is more like that of the endoplasmic reticulum than of the cell membrane itself. Novikoff and Holzman 1976, Hardin et al. 2012.

49. Novikoff and Holzman 1976, p. 80.

50. See Hardin et al. 2012.

51. The Electron microscope uses a beam of electrons that is deflected and focused by an electromagnetic field. The limit of resolution is much higher than with optical microscopy, resolving down to 0.1 nanomicrometres, with a nanomicrometre being a billionth of a metre. See Hardin et al. 2012.

52. Microtubules form a prominent part of the centrioles, which act like dancing masters during the complex manoeuvres of the chromosomes known as mitotic and meiotic cell division.

53. Gerty Theresa Cori, 1896–1957; Carl Ferdinand Cori, 1896–1984. Both received the Nobel Prize for Medicine in 1947.

54. This is now known as the Cori Cycle. Cells need energy in the form of glucose to work, repair themselves, and reproduce. The cell strips apart the glucose, releasing energy which is then stored in ATP. During anaerobic respiration, a burst of energy is produced rapidly, but this is much less than the energy produced by slower aerobic respiration. It also produces an unwanted by-product called lactate, a molecule that contains inaccessible energy. The Coris together found that this lactate is taken from muscle and transported via the blood cells to the liver. The liver then used six molecules of ATP to change two molecules of lactate into one molecule of glucose. That is the so-called Cori cycle. It results in a net loss of energy.

55. Molecules of pyruvate form an intermediate step in the chain of reactions, without the formation of lactate. Hans Adold Krebs (1900–1981) was awarded the Nobel Prize for Medicine in 1953.

56. The flagellum (plural flagella) and cilium (plural cilia) may first have been observed by Antonie van Leewenhoek in the late seventeenth century. They can be quite easily seen in *Euglena* and *Paramecium*. Some 200 years later, Swiss anatomist K. W. Zimmermann described cilia on the surface of mammalian cells for which he suggested a sensory role but this was largely ignored, and it was only during the late 1960s that the existence of sensory cilia was finally acknowledged. Since the flagellum of eukaryote cells and bacterial cells are not strictly homologous, Lynn Margulis has encouraged the term 'undulipodium' for the eukaryote cilium or flagellum. See Margulis and Dolan 2002, Margulis and Chapman 2010.

57. Bikont and Unikont are terms used by Tom Cavalier-Smith. See Cavalier Smith 2006.

58. See for example, Cavalier-Smith 2006 who places forams alongside certain kinds of algae in the 'Rhizaria'. See also useful overviews by Lecointre and Leguyader 2006, Battacharya et al. 2003, and Margulis and Chapman 2010.

59. In 1682, Nehemiah Grew first described a green precipitate within leaves. They were also seen by Antonie van Leeuwenhoek at about this time.

60. Chloroplasts can also take part in nitrogen fixation, reducing nitrate to the ammonia needed for protein synthesis. They can also reduce sulphate to hydrogen sulphide, again for protein synthesis. See Hardin et al. 2012.

61. The Calvin cycle is named after Melvin Calvin, a chemist at the University of California, who worked with colleagues on photosynthesis during the 1940s and 1950s. It was the first metabolic pathway to be worked out using the ^{14}C radioisotope. See Hardin et al. 2012.

62. When molecular biologists studied nucleic acid and protein synthesis in mitochondria and chloroplasts, they found many similarities in processes with bacterial cells. Both organelles and bacteria have circular DNA molecules without associated histones, and both show similarities in rRNA sequences, ribosome size, sensitivity to inhibitors of RNA and protein synthesis, and the type of protein factors used in protein synthesis. In addition to these molecular features, mitochondria and chloroplasts resemble bacterial cells in size and shape, and they have a double membrane in which the inner membrane has bacterial-type liquids. See Hardin et al. 2012.

63. See Woese and Fox 1977.

64. As we sailed across from Antigua to Jamaica, the astronauts of Apollo 13 came tumbling out of the sky on their ill-fated trip. As they descended, the module's cameras clicked away at the wide blue expanse of Caribbean, and our path across it. Alas, I have been unable to spot the little white speck of our two ships on these Apollo images. But it must be there somewhere.

CHAPTER 4

1. Pedro Bank was clearly shown below Jamaica by Darwin 1842 in the map from his book on Coral Reefs, though there were seemingly no reports of its coral reefs at that time. See Darwin 1842.

2. The climate of the modern world is never stable, of course. There is no such thing as *climate stasis*. Global climate change is therefore a bit of a misnomer for our present predicament—change is the norm, and has been so for aeons. But it is the magnitude of change ahead that is truly alarming. And as we shall later see in the book, things seem to have been even stranger in the deep and distant past, in the long dark ages of the Precambrian.

3. Infamous Port Royal was built upon a sand bar above a major fault zone. It was devastated by a huge earthquake on 7 June 1692 which submerged the town, leaving about 80 per cent of its population dead. Archaeologists have reputedly found a pocket watch with its hands set to 11.43 a.m.

4. Darwin in 1842 referred to the 'probable existence of submarine cliffs' and 'ledges' around various coral islands. See Darwin 1842, pp. 9, 22–3, 30. The classic paper on the formation of reef terraces during climate cycles is that by Broecker et al. 1968.

5. It may be that Pedro Bank and its neighbours also puzzled Darwin, who made oblique reference to them in 1842: 'In the West Indies there are…large and level banks, lying a little beneath the surface, ready to serve as the basis for the attachment of coral, brought occasionally into view by the entire or partial absence of reefs on them.' Darwin 1842, p. 58. He makes further allusions to this paradox of sparse coral later in the book as well (pp. 63, 65, 72).

6. *The Endeavour* was the first survey vessel to reach Australia, and to navigate coral reefs like those of the Great Barrier Reef.

7. In 2011, Tom Goreau Jr wrote to me as follows: 'My father's thesis was the first study of coral reef zonation, and the first marine ecology study done by diving'. He also reminded me that his father did 'a great deal of pioneering work on coral biochemistry, ultrastructure, and enzymology. My mother's [thesis] was on the early ontogeny and development of corals and their skeletons following settlement.'

8. Symbiosis simply means the 'living together of differently named organisms'. It comes in several varieties. Symbioses which are beneficial to both partners are called 'mutualism'; symbioses which are neither beneficial nor harmful are called 'commensalism'; and those which are harmful to one partner are called 'parasitism'. Added to this are other aspects of the relationship. Those in which one partner lives inside another are called 'endosymbionts'. Those in which one partner is photosynthetic are called 'photosymbionts'. See Smith and Douglas 1987 and Margulis 1981, 1992.

9. Researches by Robert Trench in the 1970s showed that 20–60 per cent of the products fixed by photosynthetic algae—also including maltose—were being passed on to their cnidarian hosts. The great coral reef conservationist Tom Goreau Jr is not convinced of figures much higher than this, writing to me in 2011 as follows: 'The first radiocarbon tracer experiments by my parents around 50 years ago showed that bicarbonate fixed by zooxanthellae and translocated to corals was very rapidly released as mucus and had a negligible contribution to nutrition, which is almost entirely supplied by zooplankton carnivory, just as Maurice Yonge concluded in the 1920s. My parents kept experimental reef corals alive for years completely bleached in a darkroom lab by feeding them.' See Trench 1971, 1979.

10. See Trench 1979, and Smith and Douglas 1987. The symbiont may gain not only protection from predation but also from UV irradiation. For some suggestions about nutrients and symbiont evolution, see Hallock 1988, and Brasier 1995.

11. There is much disagreement as to whether there are a single or many species of *Symbiodinium*. DNA testing shows differences between the symbionts from different corals, but the issue is whether or not these are significant enough to represent different species.

12. Their chloroplasts have chlorophylls *a* and *c*.

13. The banishment of this malarial parasite from many tropical regions is one of the great influences of the microscope. In Darwin's time, malaria was still attributed to a stench raised by the green pondweed *Chara*. See Balfour 1855, p. 579.

14. See Leggat et al. 2007.

15. See for example, Trench et al. 1981, Margulis 1981, 1992, Smith and Douglas 1987, Lee and Anderson 1991.

16. See Brasier 2009.

17. See Sollas 1905.

18. Joseph Augustine Cushman (1881–1949) was one of the first people ever to earn a living from studying Foraminifera—a line which many thousands of scientists have since followed. Next to his home he constructed a building that would become the Cushman Laboratory for Foraminiferal Research in 1923. He worked there as a consultant for oil companies, and wrote as many as 500 research papers, both aided by members of his family. Sadly, poor Cushman was dying of stomach cancer at the time of the Bikini atoll research but he pressed on heroically. His historic collections—nearly two hundred thousand microscope slides—are now housed in the famous Cushman Collection at the Smithsonian Museum in Washington. His book on Foraminifera was a classic for decades. See Cushman 1928.

19. Some key papers on atoll forams include those by Cushman, Todd, and Post 1954, and by Todd and Low 1960. Some 43 nuclear tests were fired at Enewetak between 1948 and 1958. The first hydrogen bomb was tested there in 1952, and it vaporized the island of Elugelab. Neighbouring Bikini atoll actually witnessed the most powerful American atom bomb ever exploded (15 megatonnes—a thousand times more powerful than the Hiroshima bomb). The Bravo bomb that was detonated vapourized on Bikini atoll three islands, raised water temperatures to 55,000 degrees, shook islands 200 kilometers away and left a crater 2km wide and 73m deep.

20. From the first three months spent at sea on the Pedro Bank, only 18 days proved suitable for sampling, with an average of 7 hours spent grab-sampling per month

according to my records. This caused much tension between us scientists and the Hydrographic Department.

21. Hooke described a Jurassic limestone, from northwest London in which he saw tiny pearls 'like the Cobb or Ovary of a Herring'. Their whiteness he attributed to the refraction of light beams from their surface, and in this he was correct. The open pores that he found between the ooid sand grains led him to muse that the whole Universe was made of a kind of sandy matter with a kind of liquid flowing between, and in this he was rather less correct. But it was a good try. See Hooke 1665: Observ. XV. Of Kettering-stone, and of the pores of Inanimate bodies. Schem. 9. Fig. 1.

22. For an introduction to these forams and their symbionts, see Lee and Anderson 1991.

23. See Lee and Anderson 1991.

24. In the nineteenth century, when home-spinning was more familiar, this form of shell was called 'spindle-shaped'—or fusiform in Latin—because of its resemblance to the skein of wool that winds around the spindle shaft of a spinning wheel.

25. Our problems in crossing the reef crest were not quite as severe as those faced by Charles Darwin on Cocos Keeling atoll. He punted his way into the outer reef zone with the aid of a large stick: 'I succeeded only twice in gaining this part...which I was able to reach with the aid of a leaping pole, and over which the sea broke with some violence, although the day was quite calm and the tide low...' Luckily, we did not have to pole vault across the reefs of the Pedro Cays. See Darwin 1842, p. 6.

26. The Red algae contain chlorophylls a and d as well as phycobilin proteins, red phycoerythrin (often the dominant pigment) and blue phycocyanin, carotenes, lutein, and zeaxanthin. See Hoek et al. 1995.

27. Phycoerythrin absorbs wavelengths of light in the 490–570nm spectrum. Phycocyanin absorbs light in the 550–670nm spectrum. Hence Red algae can live as deep as 268m in the Bahamas. See Thomas 2002.

28. Darwin wrote: 'The soundings...were taking with great care by Captain Fitzroy himself. He used a bell-shaped lead, having a diameter of four inches, and the armings each time were cut off and brought on board for me to examine. The arming is a preparation of tallow, placed in the concavity at the bottom of the lead. Sand, and even small fragments of rock, will adhere to it; and if the bottom be of rock it brings an exact replica of its surface.' See Darwin 1842.

29. For a student introduction to diatoms, see Brasier 1980, and Armstrong and Brasier 2004.

30. The diatoms include forms of *Nitschia* sp. See Lee and Anderson 1991.

31. Such protists provided with cilia are called ciliates and are placed in the Alveolata. Modern molecular biology is finding ciliates to be hugely abundant and diverse.

32. See Bonney 1892.

33. *Paramecium bursaria* is perhaps the most common species of freshwater protozoan to exhibit symbiosis with algae. In Europe, the species often cultivates *Micractinium* rather than *Chlorella* (Hoshina 2011). The genus *Paramecium* exhibits many other symbioses as well, including with bacteria (see Smith and Douglas 1987, and Margulis 1981, 1992, 1998).

34. Symbiosis as a term was first coined by German botanist Anton de Bary in 1873. See also Note 8 in this chapter.

35. A curious example is found in a flagellate called *Hatena arenicola*. This shows an alternation between one generation that is a free living heterotoph, feeding on other organisms, and another that cultivates a Green algal symbiont called *Nephroselmis*. See Okamoto and Inouye 2006.

36. *Paramecium bursaria* appears to be like this. According to some reports, both alga and protozoan can survive as separate entities.

37. Some invertebrates have algal symbionts which are passed on via the egg cell from the mother. Others rely on capture later in life. Some harbour whole algal cells. Others only capture algal chloroplasts. See Trench 1979, and Smith and Douglas 1987.

CHAPTER 5

1. For these associations in Tobago, we spent a happy week sampling and diving with Sally Radford.

2. See Darwin 1842, p. 61.

3. A similar phenomenon can be seen around the coast of Oman today. Along the south coast where there is strong upwelling, there are kelp forests but no reef. Along the north coast where there is no upwelling or river-runoff, the coral reefs thrive.

4. See Herschel 1830.

5. See, for example, Dobbs 2005.

6. Such as the work summarized by Broecker and Peng 1982 and Hallock 1986.

7. An ecosystem can be defined as a volume of the Earth's surface where the constituent organisms recycle energy and matter at a rate that is faster inside the system, than between it and other systems. See Botkin 1993.

8. See Hallock 1986.

9. Newspapers report many such events, such as along the coasts of China in July 2010.

10. Darwin, 1842, p. 6.

11. See Donnelly and Woodruff 2007.

12. See, for example, Baker et al. 2008.

13. See Hallock 1986, and Brasier 1995.

14. See Redfield 1934, 1958.

15. See Darwin 1859.

16. Outlines of Neo-Darwinian evolutionary theory can be found in Ridley 1997 and Mayr 2002.

17. During aerobic respiration, the Krebs cycle produces 38 molecules of ATP from one molecule of glucose. It requires freely available oxygen.

18. See Brasier 1982a, 1982b.

19. The earliest examples arguably occur at the base of the Cambrian in Canada, England, and Eastern Europe. See McIlroy et al. 2001 and Armstrong and Brasier 2004.

20. See, for example, Murray 2006.

21. See Brasier 1982a and 1982b. It was Tan Sin Hok, and later Lucas Hottinger who really started to crack this problem by looking at growth and communications from the inside out. See, for example, Hottinger 1978, 1982, 1986, 2000.

22. It was eventually published as the book 'Microfossils'. See Brasier 1980 and Armstrong and Brasier 2004.

23. Few have done more to extol the beauty of protozoan shells than Ernst Haeckel. His exquisite drawings of forams, radiolarians, acantharians, and jellyfish may have helped to inspire the spread of Art Nouveau in the later nineteenth century, as suggested by Johan Decelle and Colomban de Vargas at the 2011 ECOP protistan conference in Berlin.

24. For papers giving details of the following discussion, see Brasier 1982a, 1982b, 1984a, 1984b, 1986, 1995 and Armstrong and Brasier 2004.

25. *Archaias* cultivates the Green alga *Chlamydomonas*. *Peneroplis* cultivates the Red alga *Porphyridium*. *Sorites* cultivates the dinoflagellate *Gymnodinium*. *Borelis* cultivates the diatom *Fragilaria*. And *Heterostegina* cultivates the diatom *Nitschia*. See Lee and Anderson 1991. Indeed, the recent work of Maria Holzmann and Jan Pawlowski shows that, in

different parts of the world, these same forams always cultivate the same groups of algae. They are not at all promiscuous. See Holzmann and Pawlowski 2011.

26. See for example Brasier 1975a, 1995, and Murray 2006.

27. For architectural patterns through time, see Brasier 1982a, 1982b, 1984a. For molecular confirmation of these patterns, see Burki et al. 2010.

28. See Brasier 1984a, and Loeblich and Tappan 1988.

29. Brasier 1984a.

30. Strictly speaking, it is closer to *Ophthalmidium*, which has all of its banana-shaped chambers lying in a single plane.

31. It is a fine example of the old Haeckelian idea that 'ontogeny recapitulates phylogeny'. See Ridley 1997, pp. 184–5, and his extracts from Haeckel on pp. 211–16.

32. See, for example, Brasier 1975b.

33. See Lee and Anderson 1991; Armstrong and Brasier 2004; Murray 2006.

34. When we look at the fossil record, we will find that it was ever thus. It is the big, complex, and efficient constructions that suffer most during mass extinctions. And the simple forms sail through. There will be some important lessons for us to learn from this, in Chapter 11.

CHAPTER 6

1. This was the reef flat of the Codrington Limestone Formation, from the high stand of the last (Eemian) interglacial, some 125 thousand years ago. Sea level at that time was some 6 metres higher than now (Brasier and Donahue 1985). I have seen the same feature along the coasts of Havana and Cayo Largo in Cuba. Tom Goreau Jr tells me they can also be seen around the coasts of Yucatan, Jamaica, Bonaire, and Columbia.

2. See Brasier and Donahue 1985.

3. Lyell himself marvelled at the transport of shells in Cuba 'They may be traced even at the distance of eight or ten miles from the shore, on the summit of mountains 1200 feet high . . .' See Lyell 1850, p. 718.

4. See Watts 1990.

5. John Mather and I called this the Codrington Formation. See Brasier and Mather 1975, and Brasier and Donahue 1985.

6. John Mather and I named this the Beazer Formation. See Brasier and Mather 1975.

7. It is the Antarctic rather than the Arctic ice cap which mainly controls sea level fluctuations during glacial cycles because it has an ice cap that is largely situated up on the land rather than floating within the sea itself. A classic paper on all these ups and downs of reefs during orbitally forced (Milankovitch) climate cycles is that by Broecker et al. 1968.

8. We called this the Highlands Formation. See Brasier and Mather 1975, and Brasier and Donahue 1985.

9. This fossil alga from the Highlands Formation was only known from the Miocene of the Indo-Pacific. It was also close to *Pseudoaethesolithon* from the Miocene of Iran, as reported by Dr John Whittaker (in his Protozoa Section Report 1973/7, Natural History Museum in London). See Brasier and Mather 1975.

10. See Bartioli et al. 2005.

11. See for example Wing 2001.

12. Tom Goreau Jr tells me that the very oldest human archaeological sites in Cuba are filled with Caribbean monk seal bones. Fishermen did not really turn to turtles, queen conch, and reef fish until these the seals were wiped out.

13. Recent estimates suggest about a quarter of a million monk seals once lived in the Caribbean reefs, sustained by fish stocks vastly greater than those found in the region today. See McLennon and Cooper 2008.

14. My contact, Tom Goreau Jr, confirms that the fish now seen in Caribbean coral reefs differ from those he saw as a child before the introduction of masks, flippers, and diving gear. This is similar to a story I was hearing in Buccoo Reef in Tobago, and in Barbuda as early as 1970. Queen conch were starting to get smaller and rarer inshore. There was, it seems, a further stepwise increase in fishing stress after 1970.

15. A favourite game of Barbudans is called Worri, an old African game of skill and chance, played with shells or beans.

16. This is a development of the card game theme used within Darwin's *Lost World*. See chapter 1 in Brasier 2009.

17. Reputedly, but he may never have phrased it exactly like this.

CHAPTER 7

1. Antigua rum is pale and fragrant. Guyana rum is the kind that was once used for the naval rum tot and is typically dark and strong. There is no such thing as Barbuda rum because the island does not have sugar cane plantations.

2. Frigate birds rear their chicks for longer than any other bird. They can be as large as the adults while still in the nest.

3. In fact, an evolutionary tree is the only illustration to be found within the *Origin of Species*. See Darwin 1859.

4. Adapted from a ?sixth-century English poem called The Ruin (Gordon 1970, p. 84). This is believed to refer to the crumbling remains of the old Roman city of Aquae Sulis (modern Bath).

5. Art historian David Freedberg was scrabbling about in Windsor Castle in 1986, and came upon these drawings by Vincenzo Leonardi. These long-lost images have been taken to suggest that the earliest microscopic explorations were not undertaken by Robert Hooke back in 1665. He was simply the first man to publish them. It was among Galileo's disciples in Italy that the strange microscopic form of things was faithfully recorded. In these paintings were shown flowers, fungi, and frogs in amber. These careful drawings were later discarded or looted. Many of them were then keenly absorbed into the collections of King George III at Windsor in 1763, where they still now reside. See Freedberg 2003.

6. The idea was also explored by Reinke in 1873. See Fet and Margulis 2010.

7. About 25,000 species—a quarter of all named fungal species—are known to take part in lichen symbioses, but only two Green algae: *Trebouxia* and *Pseudotrebouxia*, as do several types of cyanobacteria. According to some estimates of the global weight of lichens, they amount to a mass great than all life in the oceans. See Margulis 1981, 1992, 1998 and Smith and Douglas 1987.

8. Fungi in the form of mycelial filaments that work symbiotically with plant roots—together called mycorrhizae—also seem very likely to have been fundamental for life's conquest of the land. See, for example, the ideas of Pirozynski and Malloch 1975, and of McMenamin and McMenamin 1994.

9. See Mereschowsky 1905.

10. There were some earlier pioneers here, too, including the German botanist A. F. W. Schimper in 1883. He had observed that chloroplasts grow and divide independently of the cells in which they reside. Indeed, he saw that they only grew from other chloroplasts. See Margulis 1981, 1992.

11. In 1927, an American biologist called Ivan Wallin from Columbia University, was to argue strongly for the role of symbiosis in evolution, but without knowledge of the work of Kozo-Polyansky. He was mainly concerned with the origin of new species through symbiosis between animals and bacteria. See Wallin 1927 and Margulis 1998.

12. For an overview, see Khakhina 1992. For a complete translation with an interesting introduction, see Fet and Margulis 2010. Among the statements of interest we can find Kozo-Polyansky saying 'Study of the symbiotic life of bacteria leads to the recognition that they participated in the origin of cells and to the

recognition of their role as cell organelles. Study of cells as such leads to the same conclusion.'

13. This technique, called molecular phylogeny, constructs evolutionary (phylogenetic) trees by mapping the sequences and divergences of genes or proteins in different organisms. The greater the divergence, then the greater the distance back to their last common ancestor. This revolution actually began in the mid 1960s with the work of Emile Zuckerkandl and Linus Pauling. A key paper is that by Woese and Fox 1977, which made use of the small subunit ribosomal RNA (or SSU rRNA). An accessible overview is that by Doolittle 2000.

14. See, for example, Margulis 1970, 1981, 1992, 1998.

15. After battles with her parents over schooling, Lynn Margulis attended college as an early entrant to take a liberal arts degree. This is where she teamed up with Carl Sagan and fell in love with science. This took place at the very start of the space race, at the time of the Russian Sputnik launch. See Margulis 1998.

16. This paper carried her first married name, Lynn Sagan (see Sagan 1967). It was then followed by up with a book (Margulis 1970).

17. It was first named the Serial Endosymbiotic Theory (or SET for short) by Max Taylor, of which the Margulis version was called the *Extreme* SET hypothesis (Taylor 1974). See also Margulis 1981, 1992 and 1998 for a discussion of its history.

18. This idea that the mitochondrion and the chloroplast evolved by a process of thinning of the cell membrane, allowing infolding, and then the pinching off of the nucleus, is known as the Direct Filiation Theory. It was put forward by Max Taylor 1974 as an alternative to the SET model. Direct evolution is still favoured by many for the origin of the flagellum and some other eukaryote structures. See Taylor 1974, Margulis 1981, 1992, 1998.

19. See Gould 1894.

20. Nor does it have Golgi bodies or endoplasmic reticulum. See Fet and Margulis 2010.

21. This later loss of the mitochondrion is suggested by the presence of relics of mitochondrial genes. For these two contrasting interpretations, see Margulis and Chapman 2010 (who place *Pelomyxa* in the Archaeprotista at the base of the tree) and Lecointre and Le Guyader 2006 (who place it near to the slime moulds).

22. It has also been said on the web that Tom likes 'to make pronouncements where others would use declarative sentences, to use declarative sentences where others would express an opinion, and to express opinions where angels would fear to tread.'

23. In 2005, I had the honour of jointly organizing a meeting on cell evolution with Tom Cavalier-Smith and T. Martin Embley, at the Royal Society in London. See Cavalier-Smith et al. 2006.

24. Later called the Archaea. See Woese et al. 1990.

25. For the evolution of these ideas, see Cavalier-Smith T. 1981, 1998, 2003, 2006, 2010.

26. Some eubacteria have also developed eukaryote-like features. *Gemmata obscuriglobus* is genetically similar to other budding bacteria. But it has the DNA enclosed within a double membrane like the nucleus; and its cell membrane lacks peptidoglycan and it can also engulf food particles. One possibility is that this is simply an example of parallel evolution with eukaryotes. But it at least suggests that the both the nuclear envelope and the phenomenon of phagocytosis could have developed within a bacterial ancestor, long before the intake of mitochondria and chloroplasts. See Fuerste and Webb 1991.

27. Note that Mereschowsky, followed by Hyman Hartman of NASA, went even further and suggested that the nucleus itself could be yet another symbiotic organelle. See Margulis 1998, p. 43.

28. Like bacteria, and unlike eukaryotes, they are reputed to respond to tetracycline but not to cyclohexamide.

29. See Maynard Smith and Szathmary 1995.

30. Intriguingly, the flow of DNA is always from symbiont to host. Where the photosymbiont is nucleated, even the nucleus may be captured and kept separately in the host, according to the recent studies of Johnson et al. 2011.

31. The splice belongs to a class of non-coding DNA called retroposons, which are frequently copied and pasted from one location to another within an animal genome. See Piskurek and Okada 2007.

32. Gene exchange seems to have taken place in the past between archaebacteria, eubacteria and eukaryotes. Many of these genes seem to have little to do with the mitochondria or chloroplasts. See Doolittle 2000.

33. Another speculation is that it resulted from the economic management of materials. Multiple copies of DNA and RNA situated within the organelles might be thought a needless expense under conditions of nutrient stress.

34. Called chaperone proteins.

35. Maynard-Smith and Szathmary 1995 dispute this, suggesting the two membranes are those of Gram Negative bacteria. This does not explain, however, the paradox of three or four layered membranes.

36. See Margulis 1981, 1992.

37. For the numerous sources for this and some of the following statements, see Delwiche 1999, Cavalier-Smith 2002, and Battacharya et al. 2003.

38. See, for example, Hoek et al. 1995.

39. See Lee and Anderson 1991.

40. The grouping of Brown phaeophytes, diatoms, and crysophytes is often known by the Cavalier-Smith term 'Chromista'.

41. Oomycete species of *Phytophthora* group are a fast evolving and economically important group. See Brasier et al. 1999.

42. See, for example, Wilcox and Wedemayer 1985. Something like this has also been reported by Johnson et al. 2011 in which a ciliate has enslaved a cryptophyte alga.

43. Algal groups thought to have chloroplasts derived from the symbiotic uptake of cyanobacteria include the chlorophytes (Green algae), the glaucocystophytes, and the rhodophytes (Red algae). These might have arisen independently. Algae with chloroplasts derived from Green algae include the chlorachniophytes and the euglenophytes. We also have at least one living organism that is seemingly arrested along this path, called *Hatena* (see Okamoto and Inouye 2006). Others with chloroplasts likely derived from Red algae include the phaeophytes (kelps, Brown algae), diatoms, and chrysophytes (often all united into the Heterokontomorpha), as well as the coccolithophorids. Those with chloroplasts possibly derived from Brown algae include the dinoflagellates. See Delwiche 1999 and Battacharya et al. 2003.

44. Tom Cavalier-Smith has suggested that symbotic uptake of a single Red algal symbiont can account for the chloroplasts within a huge grouping that he calls the Chromalveolates: the Brown algae, chrysophytes, diatoms, and Yellow Green algae. But he goes further to suggest that chloroplasts have since been lost in the forams and their relatives (the Rhizaria), as well as in the malarial parasite *Plasmodium*. This model is called the chromalveolate hypothesis (Cavalier-Smith 2002), which may be confirmed by the discovery of chloroplast genes within the nuclei of now 'naked' protozoans, such as forams and parasites.

45. Equally fascinating has been the way that these developments—about the importance of collectives and connections—have been swirling about in politics, too. Intriguingly, these paradigms have often been out of step. As the importance of the collective in biology stood its ground and began to build, so the importance of the collective in society was crumbling before the monetarist theory of capitalism. As socialist economic systems began to collapse in the political world, the importance of connections began to emerge in the world of science. Nothing ever stays put.

46. Tectis is taken from the latin word for 'building'.

47. Clasis is taken from the Greek word *klasis* for 'breaking up'.

48. And when I say class here, I mean a category that is wholly distinct at a high taxonomic level: class, division, phylum, and kingdom.

49. And more controversially, there was a fifth flip, capturing Brown algal cells.

CHAPTER 8

1. Strabo was reputedly a personal friend of the Governor of the Province of Egypt. See Jones 1932.

2. We do not know that for certain, because our sources say nothing about the time or the weather. See Jones 1932.

3. See Jones 1932, vol. 8 for the original and a translation of Strabo's Geography, book 17, chapter 1, line 34. I have made some modest changes to his choice of words.

4. They are about the size of coins. It has even been suggested that some were used as coins for barter in ancient times. See Blondeau 1972, and BouDhager-Fadel 2008.

5. In truth, it is not a single rib of rock, of course, but variable layers of limestone that were laid down during much of the Eocene. See BouDagher-Fadel 2008.

6. There are still remnants of the Tethys Ocean today, visible to us in the Aral Sea, the Caspian Sea, the Black Sea, and the Mediterranean. But these no longer form a continuous seaway.

7. According to one theory, the high temperatures of the Eocene were too warm for coral symbioses to flourish. See BouDhager-Fadel 2008.

8. When found together, I suggest that five criteria may be taken as evidence for the internal cultivation of symbiotic algae by larger benthic Foraminifera in the past. (1) restriction to a tropical to subtropical habitat that demonstrably lay within the photic zone; (2) test wall structure consistent with the passage of light through to the interior of the test (e.g. not an opaque fabric); (3) subdivision of outer chambers of the test into small box-like structures (usually called chamberlets) consistent with the housing and cultivation of photosymbionts; (4) relatively efficient routes for cytoplasmic movement through the test, as evidenced by shortened minimum lines of communication (MinLOC of Brasier 1982a; Parsimony Index (PI) of Brasier 1982b) within the shell, consistent with the energy-efficient manipulation of photosymbionts and photosynthates; (5) evolutionary trends towards gigantism, consistent with a relatively long life-span (e.g. Brasier 1988, Purton and Brasier 1999). All of these criteria are based upon features of living photosymbiotic larger benthic Foraminifera. Other features such as stable isotope fractionations have been tried (e.g. Brasier and Green 1993; Purton and Brasier 1999; Murray 2006) but are less robust and can be harder to demonstrate and interpret in the fossil record.

9. This surprising fact was first reported by Douville 1906.

10. See Blondeau 1972.

11. See note 8 above, and Ungaro 1994.

12. See Ferrandez-Canadell, C. and Serra-Kiel, J. 1992.

13. For tales of the oil industry and early life, see Brasier 2009, chapter 7.

14. See Cushman 1928.

15. See the book on *Eozoon* by Dawson 1875.

16. See for example Carpenter 1891.

17. See Lyell 1871.

18. See Darwin 1872.

19. A much fuller account of this debate can be found in Hofmann 1972, and in Brasier 2009, where an image of *Eozoon* is shown in plate 13.

20. See Kirkpatrick 1916.

21. Kirkpatrick 1916, p. 13.

22. For a full explanation of this misleading 'Mofaotyof Principle', see Brasier 2009, chapter 6.

23. Stephen Jay Gould also wrote about crazy old Kirpatrick (see Gould 1992). We shall meet some more examples of Mofaotyofs later in the book. But other embarassing examples include Thomas Henry Huxley and *Bathybius*; John Salter and *Palaeopyge*; William Carpenter and *Eozoon*; and of course myself and the Dalradian worm poo saga. See Brasier 2009.

24. See Lee and Hallock 1987, Brasier 1995, and BouDhager-Fadel 2008.

25. See Hottinger 1960.

26. For more on fusulinids, see Dunbar 1963, Ross 1992, Vachard et al. 2004, and BouDagher-Fadel 2008. For more about the biggest of all mass extinctions—at the end Permian—see books by Erwin 1993, and Benton 2003.

27. The cultivation of symbionts in brachiopod mantle tissue during the Permian was first suggested by Richard Cowen 1970.

28. The term 'mass extinction' is here used to refer to relatively short spans of geological time during which significantly raised levels of extinction took place. Conventionally there are seven such events in the last 600 million years: end Ediacaran; end Ordovician; late Devonian; late Permian; end Triassic; end Cretaceous; late Eocene. But there are an equal number of slightly less major ones: mid Ediacaran; end early Cambrian; end late Cambrian; end Jurassic; mid

Cretaceous; mid Miocene; and Holocene. And so the list goes on, with ever increasing numbers of less important events. This pattern of extinctions is now taken to show a Power Law trend, in the manner of avalanches and earthquakes: lots of little extinctions, very few big extinctions, with middling ones strung out in between. If so, this implies that both large and small extinction events may share a similar triggering mechanism. And according to the laws of Complexity, there is no expectation for a linear correlation between the magnitude of the trigger and the magnitude of the biological response. See Kauffman 1996, Bak 1997, and Buchanan 2000.

29. See Hottinger 1982, and Brasier 1986, 1995.

30. See Brasier 1986, 1995.

31. See Brasier 1988, 1995, and Hallock 1986, 1988.

32. Charles Lyell agonized about the rates of extinction at the end of a period like this one, his very own Eocene period: 'we have not yet succeeded in tracing back a perfect transition from the recent to an extinct fauna; but there are usually so many species in common to the groups which stand next in succession as to show there is no great chasm, no signs of crisis when one class of organic beings was annihilated to give place suddenly to another.' See Lyell 1850, p. 179. For more on the late Eocene calamity, see Prothero 1994, and Hallam and Wignall 1997.

33. See, for example, Service 1971, Turchin 2003.

34. See Raup 1986, and Alvarez 1997.

35. For overviews, see Raup 1991, Hallam and Wignall 1997. For some multi-author volumes in which I have also been involved, see Larwood 1988, Donovan 1989, and Walliser 1995.

36. See Winchester 2001.

37. For insights into how some of these debates worked, see Glenn 1994.

38. See Brasier 2009.

39. For the giant foram idea, see Seilacher et al. 2003 as well as its testing within Antcliffe et al. 2011. For the algal symbiont idea, see McMenamin 2000.

40. See Brasier 2009.

41. See Brasier 1989, 1995. For more ideas on reef evolution, see Wood 1999.

42. For more on the end Permian extinctions see Erwin 1993 and Benton 2003.

43. Cretaceous rudist bivalves reached several metres in diameter, and they formed major reef constructions. Their cultivation of endosymbionts seems to me plausible, as long ago suggested by Klaus Vogel 1975.

44. For such questions, see Macleod 1996, and Hallam and Wignall 1997.

45. For an overview of the evidence, see Hallam and Wignall 1997. For the role of volcanoes, see Courtillot 1999.

46. For the behaviour of complex systems, see Bak 1997, Ball 1999, 2004, and Gladwell 2000.

47. Not flawed in the sense that the rules of physics are inconstant. Those are immutable. But flawed in the sense that modern ecosystems may not be a good guide to those of the ancient past.

CHAPTER 9

1. My students Alex Liu, Leila Battison, Sean McMahon, and Kate Hibbert were studying with me in Oxford, and had joined me in Ontario. Another of my Oxford students, Nicola McLoughlin, had succeeded in getting there some years before, following in the footsteps of Stanley Awramik (e.g. Awramik and Barghoorn 1977). The aim of Nicola's expedition was to re-examine the famous stromatolites that occur in the Gunflint chert around these northern shores. That mattered to Nicola because she is a student of the biology of stromatolites. Some of the best evidence for the role of microbes in stromatolites is indeed the evidence from the Gunflint chert—where they clearly contain good microfossils. But it also mattered to Nicola because she was working with me on some non-biological ways of thinking about the ways in which stromatolites could form. On the trip reported here, we had planned to make the long trek round to the Schreiber Channel, as Stanley Tyler had done over fifty years before. Days beforehand, our reconnaissance party was prevented from getting close because of rain and slime. So, we turned adversity to advantage and got ourselves a pilot.

2. The Schreiber beach locality microfossils were first reported by Tyler and Barghoorn 1954, Barghoorn and Tyler 1965 and Cloud 1965.

3. Some resemble true lava pillows. Others are rounded cobbles from an ancient shoreline.

4. See Maliva et al. 2005.

5. Most of what follows directly relates to the original paper of Tyler and Barghoorn 1954.

6. He didn't make sense of them until he was back in Gower Street. See Browne 1995.

7. It is now known that the opening of Tutankhamun's tomb was a faked piece of theatre. Carter had privately explored the tomb in the days beforehand. See Reeves 1990.

8. See Margulis and Knoll 2005.

9. What follows differs from the story outlined in Schopf 1999. I have been greatly assisted here by discussions with Lynn Margulis and Andy Knoll, who allowed me access to their thoughts and peerless collections.

10. See Riding 2011.

11. See Brasier 2010.

12. See Hall 1883. They may have been seen by Hall in about 1876—see also Schopf 1999, p. 25.

13. He was part of an expedition from Princeton University. See Black 1933, and Jeans and Rawson 1980.

14. Including the Biwabik Formation of Minnesota. See Foster and Whitney 1851, discussed in Brasier 2010.

15. Although the term 'stromatolite' was also avoided again, by Barghoorn and Tyler (1965), the latter did mention examples of 'algal domes' from the Gunflint chert and even illustrated their own studies of modern stromatolites from Yellowstone Park. In a follow-up paper, Cloud (1965) made it clear that stromatolites do indeed occur in the Schreiber Beach region.

16. Most of my geology teachers in the 1960s knew more about coal than we wanted to hear.

17. This specimen of Michigamme coal had at first been given to Tyler's old palaeontology professor, Robert Shrock, at a meeting of the Geological Society in America at the end of 1950, who then passed it on to Barghoorn. See Margulis and Knoll 2005.

18. Originally known as Machi-gamig—this Indian name for the river in Michigan means 'the river flowing from a large lake'. Foster and Whitney 1851, p. 22.

19. Great masses of graphite had intrigued earlier generations of geologists about the possibilities of life in the Precambrian. See Dawson 1875, pp. 18, 27, and Dawson 1888, pp. 9–19.

20. It seems that the test pit for an economic coal deposit had collapsed so that no further rocks could be accessed. Tyler was therefore confined to samples scavenged from a waste tip.

21. See pp. 9–10, and Hooke 1665 (Observ. XVI. *Of* Charcoal, *or burnt* Vegetables).

22. See Hooke 1665 (Observ. XVII. *Of* Petrify'd wood, *and other* Petrify'd bodies).

23. Lyell 1850, p. 89.

24. See Seward 1917.

25. This quote, like much of the evidence in this chapter, comes from the unpublished correspondence of Barghoorn and Tyler.

26. The coals from Michigamme Slate were not reported until much later. See Tyler et al. 1957.

27. Comparable modes of fossil preservation can be seen around the geysers of Yellowstone Park today (Walter 1976) as well as around the hot springs of New Zealand (Jones et al. 2007).

28. See Kidston and Lang 1917.

29. Intriguingly, Barghoorn's first palaeobotanical publication was on the plant *Hornea* from the Rhynie chert, which he renamed *Horneophyton*. See Margulis and Knoll 2005.

30. Barghoorn and Tyler 1965.

31. Barghoorn and Tyler 1965.

32. Meaning tiny dawn star. Barghoorn and Tyler (1965) later compared it with an actinomycete bacterium.

33. According to legend, Tyler found this deposit while fishing on a Sunday. See Schopf 1999, p. 35.

34. See Margulis and Knoll 2005.

35. Barghoorn and Tyler 1965.

36. This comment comes from the unpublished correspondence of Barghoorn and Tyler.

37. This date was later refined to 1900 +/−200 million years (Barghoorn and Tyler 1965) and most recently to 1878 million years (Frallick et al. 2002).

38. See Tyler and Barghoorn 1954. In a later paper, they were being compared with the cyanobacteria *Lingbya* and *Ulothrix* (see Barghoorn and Tyler 1965). But that suggestion now seems doubtful. The *Rivularia*-like forms may be those later assigned to *Eoastrion*.

39. These and other anaerobic bacteria from modern soils were even placed within the same genus of *Kakabekia*, by Seigel and Seigel 1970. Such confident assignments have yet to be confirmed from the fossil record.

40. See Figure 9.5. Radiolaria are close relatives of the Foraminifera. These 'radiolaria' from the Gunflint chert were presumably examples of the fossil bacterium *Eoastrion*.

41. See Delwiche 1999.

42. See Huxley 1894 as well as tales of these early dredging expeditions given in Corfield 2004 and Brasier 2009.

43. See Huxley 1894.

44. First mentioned in a letter by Barghoorn to Tyler, 14 December 1953. And then in Tyler and Barghoorn 1954.

45. 'Unicellular organisms, or the dissemination of calcareous platelets of multicellular colonial forms (flagellates?) occur in considerable profusion in the chert. Of these, one distinctive form has been recognized, a cell of radially symmetrical organization, resembling that found in the genus Discoaster, a group of planktonic marine flagellates of uncertain systematic position, probably related to the Coccolithophorids.' Tyler and Barghoorn 1954, p. 607.

46. This error was corrected, albeit tacitly, in Barghoorn and Tyler 1965—they were never mentioned again.

47. Letter from Barghoorn to Tyler, 14 December 1953.

48. See Lecointre and Le Guyader 2006.

49. See Margulis and Knoll 2005.

50. That is to say, they belong to the big clan having only a single flagellum—the unikonts, and of those they belong to those with a backward pointing flagellum, like a sperm cell, called the opisthokonts. See Lecointre and Le Guyader 2006.

51. Barghoorn himself was to spend much time exploring the fungal decay of wood, and the fungal presence in the sea as well. See Margulis and Knoll 2005.

52. This assertion about the presence of fungi was tacitly corrected by Barghoorn and Tyler 1965. But is has been raised again by Krumbein 2010, in my view incorrectly.

53. See Tyler and Barghoorn (1954, figs 1 and 2). See also Cloud 1976. But tall stories about ancient fossils are always with us. See my critique (Brasier et al. 2002) of dubious cyanobacterial microfossils from the 3460 million year old Apex chert claimed by Bill Schopf (1999). Dubious as well are those cyanobacteria from outer space claimed by Richard Hoover (2011; see also the associated Commentaries).

54. Key papers that helped to change this viewpoint were those of Margulis 1970 and Woese and Fox 1977.

55. See Tyler and Barghoorn 1954.

56. See Brasier 2009. The earliest *bona fide* report of cellular preservation in Precambrian rocks seems to have been that made by Jephro Teall in the 1000 million year old Torridonian sedimentary phosphates, first reported in 1899 and soon afterwards described and illustrated; see Peach et al., 1907. A short note was included in Peat

and Diver 1982, but a part was sadly played by me in their later neglect. When I began work as a young lecturer in the Sedimentology Research Group at Reading in 1973, my neighbour along the corridor was a geologist called Sandy Stewart. For a decade, Sandy had been mapping and decoding the Torridon sediments and his work on these rocks remains without peer to this day. Sandy knew that I worked on microfossils. And he knew that I was curious about early life. For the better part of a year, he kept trying to persuade me to follow up the discoveries of old Jephro Teall. But by this time I had fallen prey to the infamous Cambrian explosion and, sadly, his entreaties fell on deaf ears. A report was at long last published by Strother et al. 2011.

57. As far back as 1922, John Gruner had described structures from the Biwabik Formation of the Mesabi mountains (Gruner 1922, plus two further papers cited in Tyler and Barghoorn 1954). Tyler wrote that 'Gruner shows pictures and describes what he believes to be algae, iron bacteria and fungi. As far as I can determine, the preservation of the materials which he had is not as good as ours'. But Cloud (1965) was much more damning about those fossils, raising the possibility of their non-biological origin and stating that new material from the same locality had revealed nothing of interest.

58. These data were reported in Barghoorn and Tyler 1965.

59. It seems strange to learn that Elso Barghoorn had not visited some key localities from a comment penned in one of Tyler's letters during January 1954. 'I noticed that you used the term "chert nodules" in the manuscript and then realized I had never mentioned to you that the plants occur in bedded cherts—with a tremendous amount of material for study. The reason I had so little material from the Schreiber locality was the long trek check there—9 or 10 miles round trip'. He would surely not have said that, if Barghoorn had been there.

60. Namely Albert Traverse, Bill Schopf, Stan Awramik, Andy Knoll, and Paul Strother. Elso Barghoorn is also said by my colleagues to have had an uncanny knack of identifying fundamental and difficult questions that urgently needed to be asked.

61. Andy Knoll did some work on the Gunflint biota and its correlatives with Barghoorn (see, for example, Knoll and Barghoorn 1976), along with Stan Awramik (Awramik and Barghoorn 1977, Knoll et al. 1978), while Paul Strother published a paper on *Huroniospora*, one of the commonest fossils found in the Gunflint chert (Strother and Tobin 1987).

62. Alas, this is true. I have heard the tapes.

63. See Margulis and Knoll 2005.

64. His researches covered a vast field. It needs to be remembered he was as much a botanist as a palaeobotanist. He worked not only on the Gunflint chert but also on

Carboniferous palaeobotany, the early geological record of flowering plants, Tertiary lignites, botanical archaeology, fungal decay, and marine mycology. He was also a great intellectual force behind the search for life in space and sought, but did not find, signs of life in the returning *Apollo* Mission samples. See Margulis and Knoll 2005.

65. He was not inactive. It seems that Barghoorn had actually amassed 800 slides of Gunflint chert by 1965. See Barghoorn and Tyler 1965.

66. See Cloud 1965.

67. Around 'Slate Island' mentioned by Cloud 1965.

68. See Cloud 1965, and Knoll 2003. Many of the fossils preserved in the Gunflint chert are now thought to be the remains of iron bacteria related to the living fossil called *Gallionella*. Hence *Gunflintia* and *Huroniospora* have also been compared by Andy Knoll with iron bacteria, while twisted branching tubes are said to resemble *Gallionella* (Knoll 2003). For a while, there was much interest in the suggestion that *Eosphaera* was a eukaryotic Green alga like modern *Volvox* (Kaźmierczak 1979) but few would accept this view now. The oldest acceptable remains of cyanobacteria are usually taken to be those described by Hans Hofmann (1976) from 2 billion year old rocks on the Belcher Islands near the eastern shore of Hudson Bay, Canada.

69. According to Bill Schopf 1999, Tyler had already written most of the geological part of this paper by 1958, and his students had taken all the fossil images too. But Barghoorn's personal problems prevented completion at that early date. Then Tyler fell ill and died, in October 1963. When the manuscript from Preston Cloud arrived for review in 1965, Barghoorn was in a state of shock. Fossil images had to be assembled by Bill Schopf from Stan Tyler's vast files without checking the fossils themselves. When the paper was finally submitted to *Science* in 1965, it was Barghoorn who took over the lead authorship, following the suggestion of Bill Schopf. When it came to the head-to-head of publication, it was clear that Cloud differed considerably from Barghoorn in his interpretation of the fossils. Although Barghoorn retreated from his earliest claims, he continued to shoehorn the remaining Gunflint fossils into familiar bacterial groups, and especially into cyanobacteria. For Preston Cloud, this approach was beginning to seem a bit blinkered. Some of those filamentous fossils, he thought, could be much better explained as iron bacteria like *Sphaerotilus* and *Gallionella*, which are both alive today in freshwater lakes. Even so, Barghoorn and Tyler (1965) did indeed compare *Gunflintia* with *Gallionella* and made reference to the common occurrence of filaments preserved in solid hematite. But they held back from taking it further.

70. See, for example, Wilson et al. 2010.

71. Brocks et al. 1999.

72. Rasmussen et al. 2008.

73. Javaux et al. 2010. While Roger Buick (2010) has raised the spectre of a eukaryote origin for these vesicles, that is doubted by Emmanuelle Javaux (personal communication, April 2011).

74. The absence of eukaryotes was elegantly revisited in Cloud (1976); see also Awramik and Barghoorn (1977) and Knoll (2003).

CHAPTER 10

1. See Lindsay 1976.

2. Those Trickledown economists were making, and are still making, a dangerous mistake, in my view. Historical studies of the biosphere may have much to tell us about what lies ahead. As we have seen, efficiency does not invite evolutionary survival. It ultimately invites extinction. Worse still, mass efficiency ultimately invites mass extinction. It is often said that those who ignore the mistakes of history are doomed to repeat them. But those who ignore the mistakes of geological history may be doomed to face extinction. That is because the biggest of lessons can arguably be read within that biggest of all history books—the fossil record itself.

3. For 1998 was an La Niña year, with extensive flooding in Queensland, allowing *Escherichia coli* to swim about unconstrained. Australia, for some reason, still has a shortage of water but no coherent strategy for water management.

4. Apollo 17 astronaut Jack Schmidt once told me he also kicked up red dust on the moon, though he was never allowed by Houston to find out why it was red.

5. Originally, the centre of the crater must have been raised up into a high mound. This had seemingly eroded away during or after the Cambrian. See Lindsay and Brasier 2006, and also Salisbury et al. 2008, who have inferred that the meteorite impact took place close to the end of the early Cambrian about 510 million years ago.

6. See Whitehouse 2001.

7. This feature had been observed and named by Ricciolus in 1655. But, as Hooke delights in telling us, the original drawings were small and crude.

8. I have corrected some of the odder spellings here. See Hooke 1665, Observ. LX. *Of the* Moon.

9. Hooke ended this section with speculations about the nature of gravity that might have set the young Isaac Newton thinking: 'To conclude therefore, it being very probable, that the Moon has a principle of gravitation, it affords an excellent distinguishing Instance in the search after the cause of gravitation, or attraction'. See Hooke 1665 (Observ. LX. *Of the* Moon).

10. These were the archaeocyathan reefs of the Lower Cambrian age (shown in Brasier 2009, fig. 2), and they died out around the globe about 510 million years ago at a time of major anoxia in shelf seas. See Zhuravlev and Wood 1996.

11. The hotel plucks its name from a song by Elvis Presley. But it earned its name from the stress suffered by the owner while trying to get it established. Business started out in the late 1970s as the Jabiru General Store, and then gradually expanded. The owner brought his bride out here for their honeymoon, but she took off with another man a few days later—hence the name.

12. This landscape was discovered by a cattle drover called Lindsay Crawford back in 1880. The local area—Cape Crawford—is named after him.

13. These shallow water sediments can be traced between the Macarthur and the Roper rivers and they are appropriately called the Roper Group.

14. In truth, it was probably in place by 4400 million years ago. See Whitehouse 2001.

15. I first learned about these Roper Group fossils while sharing a jolly dinner at the Red Lion Inn in Atherstone, UK, many years before. It was here that I first met with a lively young palaeontologist called Marjorie Muir and her students. Although she was the spouse of one of my university teachers, I never met her until I joined the quest for ancient life. By the mid 1970s, Marjorie was rather hot property in early life studies, publishing a rapid succession of papers on some very old microfossils, including some strange objects from these ancient rocks around the Lost City. By digesting these old rocks, and their associated mud-stones, within absolutely lethal concentrations of hydrofluoric acid, her student Chris Peat had been able to release some curious remains (see Peat et al. 1976). Unhappily, his portentous discoveries were never followed up in detail. But I got to know them rather well because Marjorie had decided to quit her husband and quit England too. She left for Australia. And she left me the samples from her microfossil collection, including objects from the Roper Group. A quarter of a century was to pass before these strange fossils hit the headlines again. See Javaux et al. 2001.

16. See Javaux et al. 2001, 2003, 2004, and Knoll et al. 2006.

17. The oldest *Leiosphaeridia* are about 1800 million years old and come from China. While they have been regarded as Green algal ancestors by Moczydlowska et al. 2011, they lack the surface sculptural features required here for eukaryotes.

18. *Tappania* is the oldest known fossil with external ornament. Later examples of *Tappania* have been provocatively compared with fungi by Nick Butterfield 2005 because of curious filamentous spines. This attribution seems rather doubtful, as argued by Moczydlowska et al. 2011, who regard this and the related taxa (*Shuiyousphaeridium, Trachyhystrichosphaera,* and *Dictyosphaera*) as possible Green algae

owing to their wall composition, double-wall structure and presence of an escape hole. Another suggestion is that these taxa were actually early forms of dinoflagellate resting cyst, as argued by my Chinese student Zhou Chuanming and his colleagues (see Meng et al. 2005).

19. Such as *Dictyosphaera*.

20. See Knoll et al. (2006). Although this interpretation is now widely accepted, Tom Cavalier-Smith 2006 has argued that such spines and sculpture could also have been produced by early bacteria.

21. This date of 1270 million years is not a direct one. It is said to be consistent with the local geology, and the fossil seems unlikely to be any younger than 750 million years. See Butterfield 2000.

22. Interestingly, few if any algae grew to macroscopic size until the Cambrian. According to some, *Bangiomorpha* could be reinterpreted as a kind of bacterium (Cavalier-Smith 2006).

23. An average date for the origin of eukaryotes –using molecular clocks calibrated using fossils such as these—is given as 1594 million years ago by Hedges and Kumar 2009 in their *Timetree of Life*.

24. For the earliest complex life on land, see Brasier (2009, plates 14–16) as well as Strother et al. (2011).

25. See Muir 1976.

26. For more on 'the *Conophyton* complex' of stromatolites as seen in these rocks, see Walter et al. 1988.

27. See, for example, the ideas of Paul Hoffman 1974.

28. At first our department only had one mass spectrometer. But some of my colleagues at Oxford became rather seriously hooked on mass spectrometers, and on climate change research as well. It was after this that a worrying pattern began to emerge. As climates got warmer towards the end of the millennium, so the number of mass spectrometers in our department began to multiply. By 2010, so numerous had become the mass spectrometer that our whole department was obliged to move to an entirely new and specially fitted building. At the last count, there were about a dozen mass spectrometers floating around. After a while, some of my colleagues began to worry that global warming was increasing in tandem with the increase in mass spectrometers in our department. The statistical correlation between them was strong.

29. Roger Buick in Seattle had actually called this the most boring interval in Earth history. See Buick et al. 1995.

30. See Brasier and Lindsay 1998, and Lindsay and Brasier 2000.

31. See Canfield 1998.

32. See Anbar and Knoll 2002.

CHAPTER 11

1. This can be seen in the long geological ranges of extinct taxa obtained by acid maceration of shales, including forms such *Tappania* and *Dictyosphaera*. In this respect, I agree with Bill Schopf 1999.

2. See, for example, the ideas explored by evolutionary biologists John Maynard-Smith and Eors Szathmary (1995) on major biological transitions, such as origins of life, eukaryote cells, multicellularity and tissues, and social insects. In this view, natural selection between groups (usually known as 'group selection') is seen as a historically significant evolutionary force. Traits that benefit the group as a whole may be less than advantageous for many individuals within that group. An interesting example of group selection in action is that of an insect called the water strider (*Aquarius remigis*), studied by Eldakar et al. 2010.

3. Also known as Boolean networks. In such systems, the nodes within a network can behave in an on–off manner, or in a filtered fashion. See Picton 2000.

4. There is no evidence that *Smilodon* was so specialized. But such specialisms could exist.

5. Stephen Wolfram has argued that complex systems may self-organize, or collapse, without a 'trigger'. See Wolfram 2002.

6. Fitness is a rather sloppy term that can mean different things to different people. As the American Sewell Wright long ago suggested, it may be helpful to imagine a whole range of possible fitness landscapes here. For more on this, see Ridley 1997, pp. 29–37.

7. Oxford University, Parliament, the Royal Family and the Royal Society, for example, are all at risk from too much in-breeding. Even gang-culture has its risks. Every country has institutions in constant need of refreshing its interconnections.

8. This analysis is close to that of Peter Turchin (2003). He argues that highly cooperative human societies will continue to flourish under stress, whereas those marked by divisions—that is to say, by a loss of lateral interconnections—are the ones that tend to collapse. For evolutionary ideas on why human societies cooperate, also see Henrich and Henrich 2007.

9. The biologist H. Bergson even suggested as far back as 1911 that Evolution might have realized itself within a single individual. Julian Huxley went on to speculate about, and then dismiss, these Nietzschean thoughts in 1912: 'A being to whom

accidents really could not happen might attain to that happy state through having perfected himself...by...increasing in size—until co-extensive with the universe. By...dealing with each possible eventuality....By perfecting his mental attributes....But Life, gifted with reproductive powers, has found it come cheaper and easier to choose Death for each single individual than to expend over-increasing energy on patching up the defects that are bound to appear in the individual with age.' Huxley goes on to define the individual as a heterogenous structure which shows 'Unity in Diversity'. See Huxley 1912, pp. 7–20.

10. I reconstructed these fossils and old 'Lake Beatha' towards the end of *Darwin's Lost World* in Brasier 2009. For a more detailed account, see Strother et al. 2011 and Callow et al. 2011.

REFERENCES

Allman, G.J. (1879). Address. *Report of the British Association for the Advancement of Science. 1879 Sheffield.* Pp. 1–30. John Murray, London.

Alvarez, W. (1997). *T. rex and the Crater of Doom.* Penguin, London.

Anbar, A.D. and Knoll, A.H. (2002). Proterozoic chemistry and evolution: a bioinorganic bridge. *Science,* 297, 1137–42.

Antcliffe, J.B., Gooday, A.J. and Brasier, M.D. (2011). Testing the protozoan hypothesis for Ediacaran fossils: a developmental analysis of *Palaeopascichnus. Palaeontology,* 54, 1157–75.

Araujo, A.J. (2006). The Lisbon earthquake of 1755. Public distress and political propaganda. *Journal of Portuguese History,* 4, 1–11.

Armstrong, H.A. and Brasier, M.D. (2004). '*Microfossils*'. 2nd edn. Blackwells, Oxford.

Awramik, S. and Barghoorn, E.S. (1977). The Gunflint microbiota. *Precambrian Research,* 5, 121–42.

Bak, P. (1997). *How Nature Works. The Science of Self-organized Criticality.* Oxford University Press, Oxford.

Baker, A.C., Glynn, P.W., and Riegl, B. (2008). Climate change and coral reef bleaching: an ecological assessment of long-term impacts, recovery trends and future outlook. *Estuarine, Coastal and Shelf Science,* 80, 435–71.

Balfour, J.H. (1855). *A Manual of Botany.* 3rd edn. Richard Griffin and Company, London.

Ball, P. (1999). *The Self-Made Tapestry. Pattern Formation in Nature.* Oxford University Press, Oxford.

Ball, P. (2004). *Critical Mass, How One Thing Leads to Another.* Arrow Books, London.

Barghoorn, E.S. and Tyler, S.A. (1965). Microorganisms from the Gunflint Chert. *Science,* 147, 563–77.

Bartoli, G., Sarnthein, M., Weinelt, M., Erlenkeuser, H., Garbe-Schönberg, D., and Lea, D.W. (2005). Final closure of Panama and the onset of northern hemisphere glaciation. *Earth and Planetary Science Letters,* 237, 33–44.

Battacharya, D., Hwan Su Soon, and Hackett, J.D. (2003). Photosynthetic eukaryotes unite: endosymbiosis unites the dots. *BioEssays,* 26, 50–60.

Benton, M.J. (2003). *When Life Nearly Died.* Thames & Hudson, London.

Black, M. (1933). The algal sediments of Andros Island, Bahamas. *Philosophical Transactions of the Royal Society*, Series B, 222, 165–92.

Blondeau, A. (1972). *Nummulites*. Librairie Vuibert, Paris.

Bonney, T. (1892). *The Year Book of Science*. Cassell and Company, London.

Botkin, D.B. (1993). *Forest dynamics: an ecological model*. Oxford University Press, New York.

Bottema, S. (2003). Environmental change in Crete: a 9000-year record of Holocene vegetation history and the effect of the Santorini eruption. *The Holocene*, 13, 733–49.

BouDhager-Fadel, M.K. (2008). *Evolution and Geological Significance of Larger Benthic Foraminifera*. Developments in Palaeontology and Stratigraphy. (ed. P. Wignall). Elsevier, Amsterdam.

Brasier, C.M., Cooke, D.E.L., and Duncan, J.M. (1999). Origin of a new *Phytophthora* pathogen through interspecific hybridization. *Proceedings of the National Academy of Sciences USA*, 96, 5878–83.

Brasier, M.D. (1975a). Ecology of recent sediment-dwelling and phytal Foraminifera from the lagoons around Barbuda, West Indies. *Journal of Foraminiferal Research*, 5, 193–210.

Brasier, M.D. (1975b). An outline history of seagrass communities. *Palaeontology*, 18, 681–702.

Brasier, M.D. (1980). *Microfossils*. George Allen & Unwin, London.

Brasier, M.D. (1982a). Architecture and evolution of the foraminiferid test—a theoretical approach. In Banner, F.T. and Lord, A.R. (eds) *Aspects of Micropalaeontology*. George Allen & Unwin, London, 1–41.

Brasier, M.D. (1982b). Foraminiferid architectural history: a review using the MinLOC and PI methods. *Journal of Micropalaeontology*, 1, 95–105.

Brasier, M.D. (1984a). *Discospirina* and the pattern of evolution in foraminiferid architecture. *Second International Symposium on Benthic Foraminifera, Pau, April 1983*. Elf Aquitaine, 87–90.

Brasier, M.D. (1984b). Some geometrical aspects of fusiform planispiral shape in larger foraminifera. *Journal of Micropalaeontology*, 3, 11–15.

Brasier, M.D. (1986). Form, function and evolution in benthic and planktic foraminiferid test architecture. In Leadbeater, B.S.C. and Riding, R. (eds) *Biomineralisation in Lower Plants and Animals*. Systematics Association Special Volume 30, 32–67.

Brasier, M.D. (1988). Foraminiferid extinction and ecological collapse during global biological events. In Larwood, G. P. (ed.) *Extinction and Survival in the Fossil Record*. Systematics Association Special Volume, 34, Clarendon Press, Oxford, 37–64.

Brasier, M.D. (1989). On mass extinction and faunal turnover near the end of the Precambrian. In Donovan, S.K. (ed.), *Mass Extinctions, Processes and Evidence*. Belhaven, London, 73–88.

Brasier, M.D. (1995). Fossil indicators of nutrient levels. 2: Evolution and extinction in relation to oligotrophy. In Bosence, D.W.J. and Allison, P.A. (eds) *Marine Palaeoenvironmental Analysis from Fossils*. Geological Society Special Publication, 83, 133–50.

Brasier, M.D. (2009). *Darwin's Lost World. The Hidden History of Animal Life*. Oxford University Press, Oxford.

Brasier, M.D. (2010). Towards a null hypothesis for stromatolites. In Golding, S.D. and Glikson, M. (eds) *Earliest Life on Earth. Habitats, Environments and Methods of Detection*, Springer, Dordrecht, 115–25.

Brasier, M. and Donahue, J. (1985). Barbuda—an emerging reef and lagoon complex on the edge of the Lesser Antilles Island Arc. *Journal of the Geological Society, London*, 142, 1101–17.

Brasier, M.D. and Green, O.R. (1993). Winners and losers: Recent Archaiadae from Florida Bay & Nummulites from the Eocene of Barton. *Marine Micropalaeontology*, 20, 267–76.

Brasier, M.D. and Lindsay, J.F. (1998). A billion years of environmental stability and the emergence of eukaryotes: new data from northern Australia. *Geology*, 26, 555–8.

Brasier, M.D. and Mather, J.D. (1975). The stratigraphy of Barbuda, West Indies. *Geological Magazine*, 112, 271–82.

Brasier, M.D., Green, O.R., Jephcoat, A.P., Kleppe, A.K., Van Kranendonk, M.J., Lindsay, J.F., Steele, A., and Grassineau, N.V. (2002). Questioning the evidence for Earth's oldest fossils. *Nature*, 416, 76–81.

Bright, M. (ed.) (1960a). *The Diary of Samuel Pepys. Volume Three*. J.M. Dent and Sons, London.

Bright, M. (ed.) (1960b). *The Diary of Samuel Pepys. Volume Two*. J.M. Dent and Sons, London.

Brocks, J.J., Logan, G.A., Buick, R., and Summons, R.E. (1999). Archean molecular fossils and the early rise of eukaryotes. *Science*, 285, 1033–6.

Broecker, W.S. and Peng, T.H. (1982). *Tracers in the Sea*. Eldigio Press, Lamont-Doherty Laboratory, Palisades, New York.

Broecker, W.S., Thurber, D.L., Goddard, J., Teh-lung Ku, Matthews, R.K. and Mesolella, K.J. (1968). Milankovitch Hypothesis supported by precise dating of coral reefs and deep-sea sediments. *Science*, 159, 297–300.

Browne, J. (1995). *Charles Darwin. Voyaging*. Pimlico, London.

Browne, J. (2003). *Charles Darwin. The Power of Place*. Pimlico, London.

Buchanan, M. (2000). *Ubiquity. the Science of History. Or Why the World is Simpler Than We Think*. Weidenfeld & Nicolson, New York.

Buick, R. (2010). Early life: ancient acritarchs. *Nature*, 463, 885–6.

Buick, R., Des Marais, D., and Knoll, A.H. (1995). Stable isotopic compositions of carbonates from the Mesoproterozoic Bangemall group, northwestern Australia. *Chemical Geology*, 123, 153–71.

Burki, F., Kudryavtsev, A., Matz, M.V., Aglyamova, G.V., Bulman, S., Fiers, M., Keeling, P.J., and Pawlowski. J. (2010). Evolution of Rhizaria: new insights from phylogenomic analysis of uncultivated protists. *BMC Evolutionary Biology*, 10:377, doi:10.1186/1471-2148-10-377.

Butterfield, N.J. (2000). *Bangiomorpha pubescens* n. gen., n. sp.: implications for the evolution of sex, multicellularity, and the Mesoproterozoic/Neoproterozoic radiation of eukaryotes. *Paleobiology*, 26, 386–404.

Butterfield, N.J. (2005). Probable Proterozoic fungi. *Paleobiology*, 31, 165–82.

Callow, R.H.T., Battison, L., and Brasier, M.D. (2011). Diverse microbially induced sedimentary structures from 1Ga lakes of the Diabaig Formation, northwest Scotland. *Sedimentary Geology*, 239, 117–28.

Canfield, D.E. (1998). A new model for Proterozoic ocean chemistry. *Nature*, 396, 450–3.

Carpenter, W.B. (1891). *The Microscope and its Revelations*. J. & A. Churchill, London.

Cavalier-Smith, T. (1981). Eukaryote kingdoms: seven or nine? *Biosystems*, 14, 461–81.

Cavalier-Smith, T. (1998). A revised six-kingdom system of life. *Biological Reviews*, 73, 203–66.

Cavalier-Smith, T. (2002). Chloroplast evolution: Secondary symbiogenesis and multiple losses. *Current Biology*, 12:2, R62–4.

Cavalier-Smith, T. (2003). Protist phylogeny and the high-level classification of Protozoa. *European Journal of Protistology*, 39, 338–48.

Cavalier-Smith, T.C. (2006). Cell evolution and Earth history: stasis and revolution. *Philosophical Transactions of the Royal Society*, 2361, 969–1006.

Cavalier-Smith, T. (2010). Origin of the cell nucleus, mitosis and sex: roles of intracellular coevolution. *Biology Direct*, 5:7, 1–78. [Open access, online]

Cavalier-Smith, T., Brasier, M.D., and Embley, T.M. (eds) (2006). Major steps in cell evolution: palaeontological, molecular and cellular evidence of their timing and global effects. *Philosophical Transactions of the Royal Society, Series B*, 361, 845–50.

Cloud, P.E. (1965). Significance of the Gunflint (Precambrian) microflora. *Science*, 148, 27–35.

Cloud, P.E. (1976). Beginnings of biospheric evolution and their biogeochemical consequences. *Paleobiology*, 2, 351–87.

Corfield, R. (2004). *The Silent Landscape. In the Wake of HMS Challenger 1872–1876*. John Murray, London.

Corsi, P. (2005). Before Darwin. Transformist concepts in European natural history. *Journal of the History of Biology*, 38, 67–83.

Courtillot, V. (1999). *Evolutionary Catastrophes. The Science of Mass Extinction*. Cambridge University Press, Cambridge.

Cowen, R. (1970). Analogies between the recent bivalve *Tridacna* and the fossil brachiopods Lyttoniacea and Richthofeniacea. *Palaeogeography, Palaeoclimatology, Palaeoecology*, 8, 329–44.

Craig, G.Y. (1997). *The 1785 Abstract of James Hutton's Theory of the Earth*. Edinburgh University Library, Edinburgh.

Cushman, J.A. (1928). *Foraminifera. Their classification and economic use*. Cushman Laboratory for Foraminiferal Research, Special Publication 1. Sharon, Massachusetts.

Cushman, J.A., Todd, R., and Post, R. (1954). Recent Foraminifera from the Marshall Islands. Bikini and nearby Atolls. Part 2, Oceanography (Biologic). *United States Geological Survey Professional Paper* 260-H, 319–77.

Darwin, C. (1842). *The Structure and Distribution of Coral Reefs. Being the First Part of the Geology of the Beagle under the Command of Capt. Fitzroy R.N.; during the years 1832–1836*. Smith Elder and Co., London.

Darwin, C. (1845). *Journal of Researches into the Geology and Natural History of the various countries visited by H.M.S. Beagle*. 2nd edn, John Murray, London.

Darwin, C. (1859). *On the Origin of Species by Means of Natural Selection*. 1st edn. John Murray, London.

Darwin, C. (1872). *On the Origin of Species*. 6th edn. John Murray, London.

Dawson, J.W. (1875). *Life's Dawn on Earth*. Hodder and Stoughton, London.

Dawson, J.W. (1888). *The Geological History of Plants*. Kegan Paul, Trench & Co., London.

Delwiche, C. (1999). Tracing the thread of plastid diversity through the tapestry of life. *American Naturalist*, 154. S164–S177.

Dobbs, D. (2005). *Reef Madness. Charles Darwin, Alexander Aggasiz and the Meaning of Coral*. Pantheon Books, New York.

Donnelly, J.P. and Woodruff, J.D. (2007). Intense hurricane activity over the past 5000 years controlled by El Niño and the West African Monsoon. *Nature*, 447, 465–8.

Donovan, S.K. (1989). *Mass Extinctions. Processes and Evidence*. Belhaven Press, London.

Doolittle, W.F. (2000). Uprooting the tree of life. *Scientific American*, 282, 90–5.

Douville, H. (1906). Les plus grandes Nummulites connues. *Bulletin de la Societie geologique de France*, 4:6, 81.

Drower, M. (1995). *Flinders-Petrie. A Life in Archaeology*. University of Wisconsin Press, Madison, WI.

Dunbar, C.O. (1963). Trends in evolution of American fusulines. In Von Koenigswald, G.H.R., Emeis, J.D., Buning, W.L., and Wagner, C.W. (eds) *Evolutionary Trends in Foraminifera*. Elsevier, New York, 25–44.

Dyer, B.D. (2003). *Bacteria. A Field Guide*. Cornell University Press, New York.

Eldakar, O.T., Wilson, D.S., Dlugos, M.J., and Pepper, J.W. (2010). The role of multilevel selection in the evolution of sexual conflict in the water strider *Aquarius remigis*. *Evolution*, 64, 3183–9.

Erwin, D. (1993). *The Great Paleozoic Crisis. Life and Death in the Permian*. Columbia University Press, New York.

Ferrandez-Canadell, C. and Serra-Kiel, J. (1992). Morphostructure and paleobiology of Discocyclinidae Gumbel, 1870. *Journal of Foraminiferal Research*, 22, 147–65.

Fet, V. and Margulis, L. (ed.) (2010). *Symbiogenesis. A New Principle of Evolution, by Boris Mikhaylovich Kozo-Polyansky*. Harvard University Press, Cambridge, MA.

Flannery, T. (ed.) (2001). *Terra Australis: Matthew Flinders' Great Adventures in the Circumnavigation of Australia*. The Text Publishing Company, Melbourne.

Ford, B.J. (1992). The Controversy of Robert Brown and Brownian Movement, *Biologist*, 39, 82–3.

Foster, J.W. and Whitney, J.D. (1851). *Report on the Geology and Topography of a Portion of the Lake Superior Land District in the State of Michigan. Part 11. The Iron Region, together with the General Geology*. A. Boyd Hamilton, Washington.

Frallick, P., Davis, D.W., and Kissin, S.A. (2002). The age of the Gunflint Formation, Ontario, Canada: single zircon U-Pb age determinations from reworked volcanic ash. *Canadian Journal of Earth Sciences*, 39, 1085–91.

REFERENCES

Freedberg, D. (2003). *The Eye of the Lynx: Galileo, His Friends and the Beginning of Natural History*. University of Chicago Press, Chicago, IL.

Friedrich, W.L., Kromer, B., Friedrich, M., Heinemeier, J., Pfeiffer, T., and Talamo, S. (2006). Santorini eruption radiocarbon dated to 1627–1600 B.C., *Science* 312, 548.

Fuerste, J.A. and Webb, R.I. (1991). Membrane-bounded nucleoid in the eubacterium *Gemmata obscuriglobus*. *Proceedings of the National Academy of Sciences, USA*, 88, 8184–8.

Gladwell, M. (2000). *The Tipping Point. How Little Things can make a Big Difference*. Arrow Books, London.

Glenn, W. (ed.) (1994). *Mass Extinction Debates. How Science Works in a Crisis*. Stanford University Press, Palo Alto, CA.

Gordon, R.K. (1970). *Anglo-Saxon Poetry*. Everyman's Library, Dent, London.

Gould, L. (1894). Notes on the minute structure of *Pelomyxa pelustis* (Greeff). *Quarterly Journal of Microscopical Science*, 1894, 295–305.

Gould, S.J. (1992). *The Panda's Thumb: More Reflections in Natural History*. W.W. Norton and Company. New York.

Gruner, J.W. (1922). The origin of sedimentary iron formations: the Biwabik formation of the Mesabi Range. *Economic Geology*, 17, 415, 420.

Haeckel, E. (1879). *The Evolution of Man*. C. Kegan Paul and Company, London.

Hall, J. (1883). Cryptozoon, n.g.; *Cryptozoon proliferum*, n sp. New York State Museum of Natural History, 36th Annual Report of the Trustees, New York, pp. i–vi.

Hallam, A. and Wignall, P.H. (1997). *Mass Extinctions and their Aftermath*. Oxford University Press, Oxford.

Hallock, P. (1986). Nutrient excess and the demise of coral reefs and carbonate platforms. *Palaios*, 1, 389–98.

Hallock, P. (1988). Diversification in algal symbiont-bearing foraminifera: a response to oligotrophy? In *Proceedings of the Third North American Paleontology Convention*, 1, 221–5.

Hardin, J., Bertoni, G., and Kleinsmith, L.J. (2012). *Becker's World of the Cell*. 8th edn. Pearson Benjamin Cummings, Boston, MA.

Hedges, S.B. and Kumar, S. (eds) (2009). *The Timetree of Life*. Oxford University Press, Oxford.

Henrich, J. and Henrich, N. (2007). *Why Humans Cooperate: a cultural and evolutionary explanation*. Oxford University Press, Oxford.

Herbert, S. (2005). *Charles Darwin, Geologist*. Cornell University Press, Ithaca, NY.

Herschel, J.F.W. (1830). *Preliminary Discourse on the Study of Natural Philosophy*. Longman, London.

Hoek, C. van den, Mann, D.G., and Jahns, H.M. (1995). *Algae. An Introduction*. Cambridge University Press, Cambridge.

Hoffman, P. (1974). Shallow and deepwater stromatolites in Lower Proterozoic platform-to-basin facies change, Great Slave Lake, Canada. *Bulletin of the American Association of Petroleum Geologists Bulletin*, 58, 856–67.

Hofmann, H.J. (1972). Precambrian remains in Canada: Fossils, dubiofossils, and pseudofossils. *Proceedings of the 24th International Geological Congress, Montreal*, Hapell's Press Cooperative, Gardenvale, Quebec, Section 1, 20–30.

Hofmann, H.J. (1976). Precambrian microflora, Belcher Islands, Canada: significance and systematics. *Journal of Paleontology*, 50, 1040–73.

Holzmann, M. and Pawlowski, J. (2011). Symbiosis in Foraminifera: an overview VI *European Conference on Protistology, Berlin, Germany*, Abstracts, German Society for Protozoology, Berlin, p. 71.

Hooke. R. (1665). *Micrographia. Or some Physiological descriptions of Minute Bodies made by Magnifying Glasses. With Observations and Inquiries thereupon*. Royal Society, London.

Hoover, R. (2011). Fossils of Cyanobacteria in C11 Carbonaceous Meteorites; and Commentary No 9, by Martin Brasier Life in C11 carbonaceous chondrites? *Journal of Cosmology* 13. http://journalofcosmology.com/Life100.html, last accessed 8 January 2012.

Hoshina, R. (2011). Algal symbionts of *Paramecium bursaria*. Origins and diversification. VI *European Conference on Protistology, Berlin, Germany*, Abstracts, German Society for Protozoology, Berlin, p. 72.

Hottinger, L. (1960). Uber Paleocaene und Eocaene Alveolinen. *Eclogae Geologiae Helvetiae*, 53, 265–83.

Hottinger, L. (1978). Comparative anatomy of selected foraminiferal shell structures. In R.H. Headley and G. Adams (eds) *Foraminifera III*. Academic Press, London, 203–66.

Hottinger, L. (1982). Larger foraminifera, giant cells with a historical background. *Naturwissenschaft*, 69, 361–71.

Hottinger, L. (1986). Construction, structure and function of foraminiferal shells. In Leadbeater, B.S.C. and Riding, R. (eds) *Biomineralization in Lower Plants and Animals*. The Systematics Association Special Volume, 30, 222–35.

Hottinger, L. (2000). Functional morphology of benthic foraminiferal shells: envelopes of cells beyond measure. *Micropaleontology*, 46, 37–86.

Huxley, J. (1912). *The Individual in the Animal Kingdom*. Cambridge University Press, Cambridge. Reprinted in 1995 by Oxbow Press, Woodbridge, CT.

Huxley, T.H. (1894). *Discourses: Biological and Geological. Essays by Thomas H. Huxley*. Macmillan, London.

Huxley, T.H. (1898). *Method and Results. Essays*. Appleton and Company, New York.

Jardine, L. (2004). *The Curious Life of Robert Hooke. The Man who Measured London*. Harper Perennial, London.

Jardine, L. (2009). *Going Dutch. How England Plundered Holland's Glory*. Harper Perennial, London.

Javaux, E.J., Knoll, A.H., and Walter, M. (2001). Morphological and ecological complexity in early eukaryotic ecosystems. *Nature*, 412, 66–9.

Javaux, E.J., Knoll, A.H., and Walter, M. (2003). Recognizing and interpreting the fossils of early eukaryotes. *Origins of Life and Evolution of Biospheres*, 33, 75–94.

Javaux, E.J., Knoll, A.H., and Walter, M. (2004). TEM evidence for eukaryotic diversity in mid-Proterozoic oceans. *Geobiology*, 2, 121–32.

Javaux, E., Marshall, C.P., and Bekker, A. (2010). Organic-walled microfossils in 3.2-billion-year-old shallow-marine siliciclastic deposits. *Nature* 463, 934–9.

Jeans, C. and Rawson, P.F. (1980). *Andros island, chalk and oceanic oozes—unpublished work of Maurice Black*. Yorkshire Geological Society Occasional Publication, Leicester.

Johnson, M., Brown, C., and Moeller, H. (2011). Karyoklepty and the reduced endosymbiont of *Mesodinium rubrum*, a tertiary plastid in the making? *VI European Conference on Protistology, Berlin, Germany*, Abstracts, German Society for Protozoology, p. 73.

Johnston, A. (ed.) (1965). *Francis Bacon*. B T Batsford Ltd, London.

Jones, B., de Ronde, C.E.J., and Renaut, R.W. (2007). Siliceous sublacustrine spring deposits around hydrothermal vents in Lake Taupo, New Zealand. *Journal of the Geological Society*, 164, 227–42.

Jones, H.E. (1932). *The Geography of Strabo. VIII. Book 17*. Loeb Classical Library, Harvard University Press, Cambridge, MA.

Kauffman, S. (1996). *At Home in the Universe. The Search for the Laws of Complexity*. Penguin, London.

Kaźmierczak, J. (1979). The eukaryotic nature of *Eosphaera*-like ferriferous structures from the Precambrian Gunflint Iron Formation, Canada: a compariitive study. *Precambrian Research*, 9, 1–22.

Keach, W. (ed.) (1997). *The Complete Poems. Samuel Taylor Coleridge*. Penguin, London, pp. 498–9.

Khakhina, L.N. (1992). *Concepts of Symbiogenesis. A Historical and Critical Study of the Research of the Russian Botanists*. Yale University Press, New Haven, CT.

Kidston, R.H. and Lang, W.H. (1917). On Old Red Sandstone plants showing structure from the Rhynie chert bed, Aberdeenshire. Part I. *Rhynia gwynne-vaughanii*, Kidston and Lang. *Transactions of the Royal Society of Edinburgh*, 5, 761–84.

Kirkpatrick, R. (1916). *The Nummulosphere*. III. Lamley & Co., London.

Knell, S.J. (2009). The road to Smith: how the Geological Society came to possess English geology. *Geological Society, London, Special Publication*, 317, 1–47.

Knoll, A.H. (2003). *Life on a Young Planet: The First Three Billion Years of Evolution on Earth*. Princeton University Press, Princeton, NJ.

Knoll, A.H. and Barghoorn, E.S. (1976). A Gunflint-type microbiota from the Duck Creek dolomite, Western Australia. *Origin of Life and Evolution of Biospheres*, 7, 417–23.

Knoll, A.H., Barghoorn, E.S., and Awramik, S. (1978). New microorganisms from the Aphebian Gunflint Iron Formation, Ontario. *Journal of Paleontology*, 52, 976–92.

Knoll, A.H., Javaux, E.J., Hewitt, D., and Cohen, P. (2006). Eukaryotic organisms in Proterozoic oceans. *Philosophical Transactions of the Royal Society*, 361, 1023–38.

Krumbein, W.E. (2010). Gunflint microbiota revisisted—neither stromatolites, nor cyanobacteria. In Seckbach, J. and Oren, A. (eds), *Microbial Mats, Cellular Origin, Life in Extreme Habitats and Astrobiology*. Springer, Dordrecht, 53–70.

Larwood, G.P. (ed.) (1988). *Extinction and Survival in the Fossil Record*. Systematics Association Special Volume, 34, Clarendon Press, Oxford.

Lecointre, G. and Le Guyader, H. (2006). *The Tree of Life. A Phylogenetic Classification*. The Belknap Press, Harvard University, Cambridge, MA.

Lee, J.J. and Anderson, R.O. (1991). *Biology of Foraminifera*. Academic Press, London.

Lee, J.J. and Hallock, P. (1987). Algal symbioses as the driving force in the evolution of larger foraminifera. *Annals of the New York Academy of Sciences*, 503, 330–47.

Leggat, W., Hoegh-Guldberg, O., Dove, S., and Yellowlees, D. (2007). Analysis of an EST library from the dinoflagellate (*Symbiodinium* sp.) symbiont of reef-building corals. *Journal of Phycology*, 43, 1010–21.

Lindsay, J.F. (1976). *Lunar stratigraphy and sedimentology*. Developments in Solar System- and Space Science, No. 3. Elsevier Scientific Publishing Co., Amsterdam.

Lindsay, J.F. and Brasier, M.D. (2000). A carbon isotope reference curve for ca. 1700–1575 Ma, McArthur and Mount Isa Basins, Northern Australia. *Precambrian Research*, 99, 271–308.

Lindsay, J. and Brasier, M. (2006). Impact structures as biospheric microenvironments: Lawn Hill Structure, Northern Australia. *Astrobiology*, 6, 346–63.

Lister, J.J. (1830). On some properties in achromatic object glasses applicable to the improvement of the microscope. *Philosophical Transactions of the Royal Society*, 120, 187–200.

Litchfield, H.E. (ed.) (1904). *Emma Darwin, wife of Charles Darwin: a century of family letters*. Volume 1, Cambridge, privately printed.

Loeblich, A.R. and Tappan, H. (1988). Foraminiferal evolution, diversification and extinction. *Journal of Palaeontology*, 62, 695–714.

Lyell, C. (1832). *Principles of Geology*. 2nd edn. John Murray, London.

Lyell, C. (1837). *Principles of Geology*. 5th edn. John Murray, London.

Lyell, C. (1850). *Principles of Geology*. 8th edn. John Murray, London.

Lyell, C. (1871). *The Student's Elements of Geology*. John Murray, London.

Mabberley, D.J. (1985). *Jupiter botanicus: Robert Brown of the British Museum*. Lubrecht and Cramer Ltd, Port Jervis, NY.

McCalman, I. (2010). *Darwin's Armada*. Penguin, London.

McIntyre, D.B. and McKirdy, A. (1997). *James Hutton. The Founder of Modern Geology*. The Stationery Office, Edinburgh.

McIlroy, D., Green, O.R., and Brasier, M.D. (2001). Palaeobiology and evolution of the earliest agglutinated Foraminifera: *Platysolenites*, *Spirosolenites* and related forms. *Lethaia*, 34, 13–29.

McKnown, D.B. (2003). *Behold the Antichrist: Bentham on Religion*. Prometheus Books, Amherst, NY.

McLennon, L. and Cooper, A.B. (2008). Extinction rate, historical population structure and ecological role of the Caribbean monk seal. *Proceedings of the Royal Society, Series B*, 275, 1351–8.

MacLeod, N. (1996). K-T redux. *Paleobiology*, 22, 311–17.

McMenamin, M.A. and McMenamin, D.S. (1994). *Hypersea. Life on Land*. Columbia University Press, New York.

McMenamin, M.A.S. (1998). *The Garden of Ediacara*. Columbia University Press, New York.

McNeil, H. (1997). *Emily Dickinson*. Everyman Library, London.

Maliva, R.G., Knoll, A.H., and Simonson, B.M. (2005). Secular changes in the Precambrian silica cycle: insights from chert petrology. *GSA Bulletin*, 117, 835–45.

Margulis, L. (1970). *Origin of Eukaryotic Cells*. Yale University Press, New Haven, CT.

Margulis, L. (1981). *Symbiosis in Cell Evolution*. Freeman and Company, San Francisco, CA.

Margulis, L. (1992). *Symbiosis in Cell Evolution. Microbial evolution in the Archaean and Proterozoic eons*. Freeman, New York.

Margulis, L. (1998). *The Symbiotic Planet. A New Look at Evolution*. Weidenfeld & Nicolson, London.

Margulis, L. and Chapman, M.J. (2010). *Kingdoms and Domains. An Illustrated Guide to the Phyla of Life on Earth*. Academic Press, Amsterdam.

Margulis, L. and Dolan, M.F. (2002). *Early Life. Evolution on the Precambrian Earth*. Jones and Bartlett, Sudbury, MA.

Margulis, L. and Knoll, A.H. (2005). Elso Sterrenberg Barghoorn Jr, 1915–1984. *Biographical Memoirs*, 87, 1–19. The National Academies Press, New York.

Maynard Smith, J. and Szathmary, E. (1995). *The Major Transitions in Evolution*. Oxford University Press, Oxford.

Mayr, E. (2002). *What Evolution Is*. Weidenfeld and Nicolson, London.

Meng, E.W., Zhou, C.M., Yin, L.M., Chen, Z.L., and Yuan Z.L. (2005). The oldest known dinoflagellates: morphological and molecular evidence from Mesoproterozoic rocks at Yongji, Shanxi Province. *Chinese Scientific Bulletin*, 50, 1230–4.

Mereschowsky, C. (1905). Über Natur und Ursprung der Chromatophoren im Pflanzenreiche. *Biologisches Zentralblatt*, 25, 593–604.

Moczydlowska, M., Landing, E., Zang, W., and Palacios, T. (2011). Proterozoic phytoplankton and timing of chlorophyte algae origins. *Palaeontology*, 54, 721–33.

Muir, M.D. (1976). Proterozoic microfossils from the Amelia Dolomite, McArthur Basin, Northern Territory. *Alcheringa*, 1, 143–58.

Muir, M.D. (1983). Proterozoic microfossils from the Mara Dolomite Member, Emmerugga Dolomite, McArthur Group, from the Northern Territory, Australia. *Botanical Journal of the Linnean Society*, 86, 1–18.

Murray, J. W. (2006). *Ecology and Applications of Benthic Foraminifera*. Cambridge University Press, Cambridge.

Novikoff, A.B. and Holzman, E. (1976). *Cells and Cell Organelles*. Holt, Rinehart and Winston, New York.

Okamoto, N. and Inouye, I. (2006). *Hatena arenicola* gen. et sp. nov., a Katablepharid undergoing probable plastid acquisition. *Protist*, 157, 401–19.

Palm-Gold, J. (2011). The Rookery. Exploring London's worst slum. *Current Archaeology*, 258, 26–33.

Peach, B.N., Horne, J., Gunn, W., Clough, C.T., Hinxman, L.W., and Teall, J.J.H. (1907). *The Geological Structure of the North-West Highlands of Scotland*. Memoirs of the Geological Survey of Great Britain, Edinburgh.

Peat, C. and Diver, W. (1982). First signs of life on Earth. *New Scientist*, 95, 776–8.

Peat, C.R., Muir, M.D., Plumb, K.A., McKirdy, D.M., and Norvick, M.S. (1976). Proterozoic microfossils from the Roper Group, Northern Territory, Australia *Journal of Australian Geology and Geophysics*, 3, 1–17.

Picton, P.D. (2000). *Neural Networks*. 2nd edn. Palgrave Macmillan, London.

Pirozynski, K.A. and Malloch, D.W. (1975). The origin of land plants: A matter of myco-trophism. *Biosystems*, 6, 153–64.

Piskurek, O. and Okada, N. (2007). Poxviruses as possible vectors for horizontal trans-fer of retroposons from reptiles to mammals. *Proceedings of the National Academy of Sciences, USA*, 104, 12046–51.

Prothero, D.R. (1994). *The Eocene–Oligocene Transition*. Columbia University Press, New York.

Purton, L.M.A. and Brasier, M.D. (1999). Giant protist *Nummulites* and its Eocene envi-ronment: life span and habitat insights from $\delta^{18}O$ and $\delta^{13}C$ data from *Nummulites* and *Venericardia*, Hampshire basin, UK. *Geology*, 27, 711–14.

Rasmussen, B., Fletcher, I.R., Brocks, J.J., and Kilburn, M.R. (2008). Reassessing the first appearance of eukaryotes and cyanobacteria. *Nature*, 455, 1101–4.

Raup. D.M. (1986). *The Nemesis Affair. A Story of Dinosaurs and the Ways of Science*. W.W. Norton and Company, New York.

Raup. D.M. (1991). *Extinction. Bad Genes or Bad Luck?* W.W. Norton and Company, New York.

Raven, J.A. and twelve others, (2002). Seaweeds in cold seas: evolution and carbon ac-quisition. *Annals of Botany* 90, 525–36.

Redfield A.C. (1934). On the proportions of organic derivations in sea water and their relation to the composition of plankton. In Daniel, R.J. (ed.) *James Johnstone Memorial Volume*. University Press of Liverpool, Liverpool, 177–92.

Redfield, A.C. (1958). The biological control of chemical factors in the environment, *American Scientist*, 64, 205–21.

Reeves, C. N. (1990). *Valley of the Kings: the decline of a royal necropolis*. Kegan Paul, London.

Riding, R. (2011). The nature of stromatolites: 3,500 million years of history and a cen-tury of research. In Reitner, J., Queric, N.-V. and Arp, G. (eds) *Advances in Stromatolite Geobiology*. Lecture Notes in Earth Sciences. Springer, Dordrecht, 29–74.

Ridley, M. (1997). *Evolution*. 2nd edn. Oxford University Press, Oxford.

Ross, C.A. (1992). Paleobiogeography of Fusulinacean Foraminifera. Studies in Benthic Foraminifera. *Proceedings of the Fourth International Symposium on Benthic Foraminifera, Sendai, 1990. Tokyo*. Tokai University Press, Tokyo, 22–31.

Sagan, L. (Lynn Margulis) (1967). On the origin of mitosing in cells. *Journal of Theoretical Biology*, 14, 225–75.

Salisbury, J.A., Tomkins, A.G., and Schaefer, B.F. (2008). New insights into the size and timing of the Lawn Hill impact structure: relationship to the Century Zn-Pb de-posit. *Australian Journal of Earth Sciences*, 55, 587–603.

Schopf, J. W. (1999). *Cradle of Life: The Discovery of Earth's Earliest Fossils*. Princeton Univer-sity Press, Princeton, NJ.

Seigel, B.Z. and Seigel, S.M. (1970). Biology of the Precambrian genus *Kakabekia*: New observations on living *Kakabekia barghoorniana*. *Proceedings of the National Academy of Sciences USA*, 67, 1005–10.

Seilacher, A., Grazhdankin, D., and Legouta, A. (2003). Ediacara biota: the dawn of animal life in the shadow of giant protists. *Paleontological Research*, 7, 43–54.

Service, E.R. (1971). *Cultural Evolutionism: Theory in Practice*. R. & W. Holt, London.

Seward, A.C. (1917). *Fossil Plants. A Text-Book for Students of Botany and Geology*: 4 Volumes (published from 1898–1917). Cambridge University Press, Cambridge.

Smith, C.H. and Beccaloni, G. (2010). *Natural Selection and Beyond, The Intellectual Legacy of Alfred Russel Wallace*. Oxford University Press, Oxford.

Smith, D.C. and Douglas, A.E. (1987). *The Biology of Symbiosis*. Edward Arnold, London.

Sollas, W.J. (1905). *The Age of the Earth and Other Geological Studies*. T. Fisher Unwin, London.

Stewart, G.B. (2003). *Microscopes*. The Kid Haven Science Library, Kid Haven Press, Farmington Hills, MI.

Strother, P.K. and Tobin, K. (1987). Observations on the genus *Huroniospora* Barghoorn: implications for paleoecology of the Gunflint microbiota. *Precambrian Research*, 36, 323–33.

Strother, P.K., Battison, L., Brasier, M.D., and Wellman, C.H. (2011). Earth's earliest non-marine eukaryotes. *Nature*, 473, 505–9.

Taylor, F.J.R. (1974). Implications and extensions of serial endosymbiotic theory of the origin of eukaryotes, *Taxon*, 23, 229–58.

Thomas, D. (2002). *Seaweeds*. The Natural History Museum, London.

Todd, R. and Low, D. (1960). Smaller Foraminifera from Eniwetok Drill Holes. Bikini and nearby Atolls, Marshall Islands. *United States Geological Survey Professional Paper* 260-X, 799–861.

Trench, R.K. (1971). The physiology and biochemistry of zooxanthellae symbiotic with marine coelenterates. I. The assimilation of photosynthetic products of zooxanthellae by two marine coelenterates. *Proceedings of the Royal Society of London, Series B*, 177, 225–35.

Trench, R.K. (1979). The cell biology of plant-animal symbioses. *Annual Reviews of Plant Physiology*, 30, 485–531.

Trench, R.K., Wethey, D.S., and Porter, J.W. (1981). Observations on the symbiosis with zooxanthellae among the Tridacnidae (Mollusca, Bivalvia). *Biological Bulletin*, 161, 180–98.

Turchin, P. (2003). *Historical Dynamics: Why States Rise and Fall*. Princeton Studies in Complexity. Princeton University Press, New York.

Tyler, S.A. and Barghoorn, E.S. (1954). Occurrence of structurally preserved plants in Pre-Cambrian rocks of the Canadian Shield. *Science*, 119, 606–8.

Tyler, S.A., Barghoorn, E.S., and Barrat, L.P. (1957). Anthracitic coal from Precambrian upper Huronian black shale of the iron river district, northern Michigan. *Bulletin of the Geological Society of America*, 68, 1293–304.

Ungaro, S. (1994). Nummulite morphological evolution. In Matteucci R., Carboni, M.G., and Pignatti, J.S. (eds) *Studies on Ecology and Paleoecology of Benthic Communities. Bolletino della Societa Paleontologica Italiana*, Special Volume, 2, 343–9.

Vachard, D., Munnecke, A., and Servais, T. (2004). New SEM observations of keriothecal walls: implications for the evolution of the Fusulinida. *Journal of Foraminiferal Research*, 34, 232–42.

Vasil, I.K. (2008). A history of plant biotechnology: from the Cell Theory of Schleiden and Schwann to biotech crops. *Plant Cell Reports*, 27, 1423–40.

Vogel, K. (1975). Endosymbiotic algae in Rudists. *Palaeogeography, Palaeoclimatology, Palaeoecology*, 17, 327–32.

Wallin, I.E. (1927). *Symbioticism and the Origin of Species*. Williams and Wilkins, Baltimore.

Walliser. O. (ed.) (1995). *Global Events and Event Stratigraphy in the Phanerozoic*. Springer, Berlin.

Walter, M.R. (1976). Hot spring sediments in Yellowstone National Park. In Walter, M.R. (ed.) *Stromatolites. Developments in Sedimentology*, 20, 489–98.

Walter, M.R., Krylov, I.N., and Muir, M.D. (1988). Stromatolites from the Middle and Late Proterozoic sequences in the McArthur and Georgina Basins and the Mount Isa Province, Australia. *Alcheringa: An Australasian Journal of Palaeontology*, 12, 79–106.

Watts, David. (1990). *The West Indies: Patterns of Development, Culture, and Environmental Change Since 1492*. Cambridge University Press, Cambridge.

White, M. (1998). *Isaac Newton. The Last Sorcerer*. Fourth Estate, London.

Whitehouse, D. (2001). *The Moon: a Biography*. Headline Book Publishing, London.

Wilcox, L.W. and Wedemayer, G.J. (1985). Dinoflagellate with blue-green chloroplasts derived from an endosymbiotic eukaryote *Science*, 227, 192–4.

Williams, R.J.P. and Frausto da Silva, J.J.R. (1996). *The Natural Selection of the Chemical Elements*. Oxford University Press, Oxford.

Wilson, J.P. and seventeen others. (2010). Geobiology of the late Paleoproterozoic Duck Creek Formation, Western Australia. *Precambrian Research*, 179, 135–49.

Winchester, S. (2001). *The Map that Changed the World. William Smith and the Birth of Modern Geology*. HarperCollins, London.

Wing, E.S. (2001). The sustainability of resources used by Native Americans on four Caribbean islands. *International Journal of Osteoarchaeology*, 11, 112–26.

Woese, C. and Fox, G. (1977). Phylogenetic structure of the prokaryotic domain: the primary kingdoms. *Proceedings of the National Academy of Sciences*, 74: 5088–90.

Woese, C., Kandler, O., and Wheelis, M.L. (1990). Towards a natural system of organisms: proposal for the domains Archaea, Bacteria, and Eucarya. *Proceedings of the National Academy of Sciences, USA*, 87, 4576–9.

Wolfram, S. (2002). *A New Kind of Science*. Wolfram Media, Champaign, IL.

Wood, R. (1999). *Reef Evolution*. Oxford University Press, Oxford.

Zhuravlev, A.Yu. and Wood, R. (1996). Anoxia as the cause of the mid-Early Cambrian (Botomian) extinction event. *Geology*, 24, 311–14.

INDEX